IMAGES
of America

EARLY WHITEWATER INDUSTRY

On the cover: This photograph, taken around 1900, shows a collection of Whitewater farmers of all ages. In the beginning of the 20th century, farmers began to cooperate in new ways. They began to work together in production and marketing, improving both the quality of their product and their profits. They also began to purchase and share traction engines, like the one pictured here. These steam-powered machines originally had to be pulled by a horse, as they were impossible to steer. However, once these machines featured steering wheels, they became much easier to use for plowing, pulling, and transportation. (Whitewater Historical Society.)

IMAGES
of America

EARLY WHITEWATER INDUSTRY

Bo McCready

ARCADIA
PUBLISHING

ISBN 978-0-7385-6189-9

Published by Arcadia Publishing
Charleston SC, Chicago IL, Portsmouth NH, San Francisco CA

Printed in the United States of America

Library of Congress Catalog Card Number: 2008929807

For all general information contact Arcadia Publishing at:
Telephone 843-853-2070
Fax 843-853-0044
E-mail sales@arcadiapublishing.com
For customer service and orders:
Toll-Free 1-888-313-2665

Visit us on the Internet at www.arcadiapublishing.com

To Brittany, my best friend

Contents

ACKNOWLEDGMENTS

First and foremost, I would like to thank Prof. Colleen Dunlavy at the University of Wisconsin for her advice, oversight, and support. From the day that I began this project, she made sure that I kept an eye on the bigger picture and helped me develop as a writer and historian. Prof. James Donnelly was also an invaluable contributor whose encouragement kept my spirits up when I was buried in primary source material, and Prof. Michael Chamberlain consistently made sure that we history majors did not lose our sanity as we finished our final projects.

Many thanks are also due to many residents of Whitewater, especially Carol Cartwright of the Whitewater Historical Society (WHS) for her incredible patience and assistance, Shirley Hapka and the Irvin L. Young Memorial Library for their support and essential resources, and Fred Kraege, whose many years of research and compilation of Whitewater's history made this project possible. And many thanks to Anne Gray McCready, who spent numerous hours in her tireless efforts to collect and scan photographs from many different sources, allowing me to focus on research and writing, making my job as an author much easier. Thank you to all who contributed photographs for this book.

Some of this research would not have been possible without the financial support of Tricia Nordby Hamrin and the Letters and Science Honors Program at the University of Wisconsin. Finally, I must thank my parents for their unconditional support, as well as all other friends and family.

INTRODUCTION

By all appearances, the town of Whitewater, Wisconsin, grew in the same manner as many other midwestern communities. Whitewater was settled in 1837 as a farming community, but it was the coming of a land-grant railroad in 1852, the Milwaukee and Mississippi, that allowed Whitewater to prosper, as the town's population doubled in three years following the railroad's arrival. A myriad of new inventions and patents came from Whitewater's ingenuous business population. Its manufacturing leaders developed not only their own variations on existing products but also completely new machines and processes. Whitewater's farmers and dairymen consistently won prizes for their products, while its factories produced reapers and wagons that won national competitions. Whitewater steadily grew into a bustling manufacturing center during the 19th century. This was truly a town that showed remarkable potential.

Whitewater was not just another small village. Rather, it featured two factories that achieved national fame. The Esterly Harvesting Machine Company produced a reaper to rival that of Cyrus Hall McCormick, who produced the famous Virginia Reaper. Esterly and McCormick enjoyed a contentious relationship that stemmed from Esterly's questionable victory over McCormick at an 1848 reaper trial in Chicago. The Esterly reaper sold well throughout the West, and it was very prominent on the wheat frontier. A small town in South Dakota even called itself "Esterly" due to its reliance on the reaper, although the town has since disappeared.

Meanwhile, the Whitewater Wagon produced by the Winchester and Partridge Manufacturing Company achieved its own success. The wagon won many prizes, including one at the Philadelphia Centennial Exposition. Queen Victoria herself selected the Whitewater Wagon for a border-making expedition in Canada. Whitewater was by no means a huge manufacturing power, but its factories achieved their own national prominence.

Towns like Whitewater contributed immeasurably to the development of industry in the Midwest. Fifty percent of all industrial workers in the Midwest worked in small towns, thus to only focus on the large cities fails to tell half the story. Chicago may have had the McCormick reaper factory, but Whitewater had the Esterly reaper factory, and while its capacity was much smaller, it was still a substantial contributor to Wisconsin's economic health.

Then, in the 1890s, the town's development suddenly stopped. This developing manufacturing town lost its economic clout in a few short years. In 1890, Whitewater's population stood at 4,359. By 1905, the town claimed 3,108 residents. What exactly happened in Whitewater? Why did the town's steady growth give way to a precipitous decline? This collapse has gone unexplained for the last 100 years.

Upon first glance, the panic of 1893 serves as an easy explanation for the town's struggles. The panic caused widespread unemployment and the near-collapse of the American dollar. The nation only recovered after substantial government action. Existing scholarship invariably deals with bank collapses, the Pullman strikes, the free silver question, and Jacob Coxey's army of unemployed and their march to Washington in 1894.

Economic, social, and political historians alike have thoroughly dissected 1893, its causes, and its general effects. However, scholars have rarely addressed the effects of the panic on individual towns. What happened in the towns that lost the most from 1890 to 1895? Why did some towns lose so much while others continued, seemingly unaffected? More importantly, could Whitewater and other small towns have successfully mitigated the effects of the panic?

While many towns failed during the 1890s, cities as few as 15 miles away from Whitewater prospered. Larger economic trends provide a framework in which to view Whitewater, but the town itself was a product of unique circumstances and events. The loss of the town's most important businesses significantly retarded Whitewater's growth, as no industries quickly emerged to replace them. Even in the face of urgent pleas from its business leaders, Whitewater's local government did little to lure new industries into the town. Numerous proposed railroad lines also promised to come to Whitewater and then failed to materialize.

If the panic of 1893 cannot fully explain what happened to Whitewater, then what does explain it? Perhaps one might consider the existing literature on deindustrialization as a source for a possible alternate explanation. However, existing deindustrialization literature does not satisfactorily explain the town's decline. To begin, deindustrialization in 19th-century America has been almost completely ignored by existing scholarship. Only a few historians actually point out that the history of American deindustrialization actually begins in the second half of the 19th century, rather than during the 1920s.

There are a vast number of scholarly journal articles that have been written on deindustrialization, but these articles overwhelmingly describe deindustrialization as a 20th-century postwar phenomenon concentrated in the Rust Belt. Recent scholarship frames deindustrialization as a consequence of globalization and foreign direct investment, two factors that are clearly not at play in this study. Studies on 20th-century deindustrialization should encourage historians to look further back, but few have done so. Nineteenth-century deindustrialization in the United States has been largely ignored, especially regarding communities in the Midwest.

What the deindustrialization literature does, however, is encourage the study of different cases through a combination of methods, including analysis of business, economic, political, and labor history. Whitewater provides a historian with an opportunity to study deindustrialization in an often-ignored time frame and location.

Scholars have also rarely addressed the effects of economic decline on individual small towns. It was towns like Whitewater that contributed immeasurably to the development of industry in the Midwest. Not every town was a Chicago or Milwaukee, and these larger cities did not completely dominate Midwestern manufacturing.

The case of Whitewater is an anomaly that must be understood, not only for the sake of the town and state, but for the sake of other small towns that may wish to either avoid or explain their own struggles. Chapter 5 describes some of the most notable businesses in Whitewater during the last years of the 19th century, a period in which the fate of the town hung in the balance.

With the loss of its major industries, there was little incentive for the common worker to stay in Whitewater. There was also little incentive for another factory to move to Whitewater, particularly in the wake of the panic of 1893. More than 15,000 business and 600 banks failed in the year 1893 alone. The roots of the panic began three years before, after the near failure of the Baring Brothers Bank of London. The "London Crash" led anxious English investors to withdraw their capital from eastern American markets and banks. What could have been a smaller recession became a full-fledged depression, as a run on the government's gold reserves threatened the stability of the American dollar.

Whitewater's economic failure was the result of many different factors, both avoidable and unavoidable. It resulted from the indifference of local leaders, ignorance of national trends, and a populace that failed to recognize the problems that were right in front of them. But it also resulted from national and regional economic conditions, labor strikes, and the success of the town's competitors.

There is no mention in Whitewater of business incentives or any response to the panic of 1893. This is especially troubling when placed in contrast to an address by Oshkosh mayor Arthur Kellogg in 1895: "Invite new industries to locate here, protect those we have, guard closely the people's interest if you are asked to grant a franchise, for whatever it may be, and at the same time see that the privileges and stipulations granted in the franchise are lived up to." This is a far cry from the short-sighted plans of some of Whitewater's mayors.

Whitewater was also the victim of bad luck and unfortunate timing. Whitewater's two major factories left in the three years before the panic of 1893. Perhaps if they had left earlier, Whitewater would have had time to acquire another significant factory before economic conditions worsened to the point that this was impossible. If the factories had stayed in Whitewater longer, perhaps the town would have rallied around its major manufacturers even in the face of financial hardship and continued its development as a manufacturing center. It seems that even fate had its way with Whitewater.

Based on the far-reaching nature of the panic of 1893, it is unlikely that Whitewater could have avoided its effects. However, questionable decisions within the town ensured that Whitewater sunk even deeper into its local recession. Years of deficit spending finally caught up with the town, yet this spending pattern did not stop. Thousands of dollars that could have been used to prop up Whitewater's businesses were instead spent on sidewalks and other seemingly nonessential public projects. When factory closures combined with an intercontinental depression, the town was simply unable to maintain its economic output. In hindsight, it is easy to see what went wrong in Whitewater. But Whitewater was just one of thousands of towns crushed by 1893. The citizens and policy makers of Whitewater made some poor decisions, but they were not alone. If Whitewater had kept its major factories, would the end result have been the same? It is certainly possible. The Esterly Company could have continued its steady operations, rather than losing all of its capital in patent lawsuits, and Winchester and Partridge could have tapped into new markets with their superior wagons, but the leaders of these companies all died before the onset of the panic. Would their Whitewater-based successors have had the necessary acumen to weather the financial storm?

This book contains the story of Whitewater, a booming midwestern town that suddenly collapsed. These pages tell of the individuals who made the town a success, but they also explain why Whitewater's success did not continue. Indeed, the story of Whitewater proves that even after a town's largest factories are gone, all is not lost, as demonstrated in chapter 6. Whitewater is a fascinating case study in urban growth and sudden deindustrialization, as well as economic recovery, and the issues and events discussed in this introduction are of great national significance as well. What follows is the story of a remarkable town.

One

Whitewater's Early Years

The story of Whitewater is similar to that of many other midwestern towns, and this section will give historians and readers alike greater insight into the development of a typical Wisconsin community. But Whitewater also had its own unique features, such as Wisconsin's first paper mill and some of the state's largest early factories. This was truly a pioneer town that achieved success even before Wisconsin's incorporation as a state.

Whitewater eventually became a thriving manufacturing town with hundreds of factory workers. But it took many years to build Whitewater into a manufacturing hotbed. Early Whitewater was built around many smaller industries that provided the town with necessary services and employment, helping Whitewater grow. However, the town was not just a simple farm community. It was a diverse, successful town, and its early citizens were waiting for the manufacturing establishments that would make the town even stronger. This chapter outlines the early development of Whitewater.

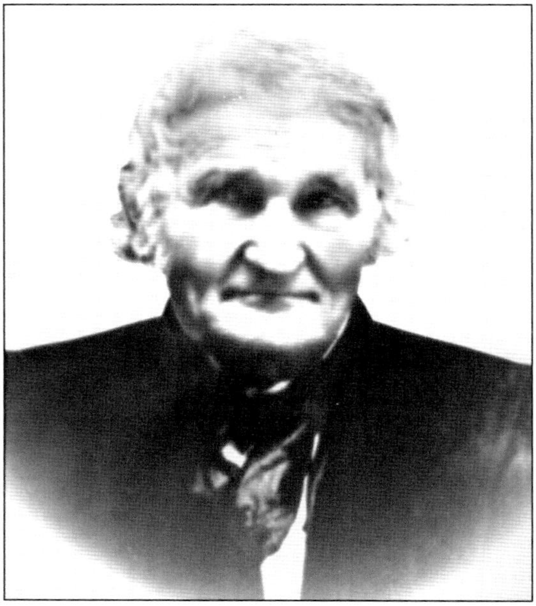

Samuel Prince built the first cabin in Whitewater in 1837. In April 1837, he made his claim by marking his name and date on a tree. He then traveled to Milton, Janesville, Delavan, and Troy to make sure Whitewater was the right choice. By July, he was satisfied and began construction on a log cabin measuring 12 feet by 12 feet, with an 8-foot ceiling. (WHS.)

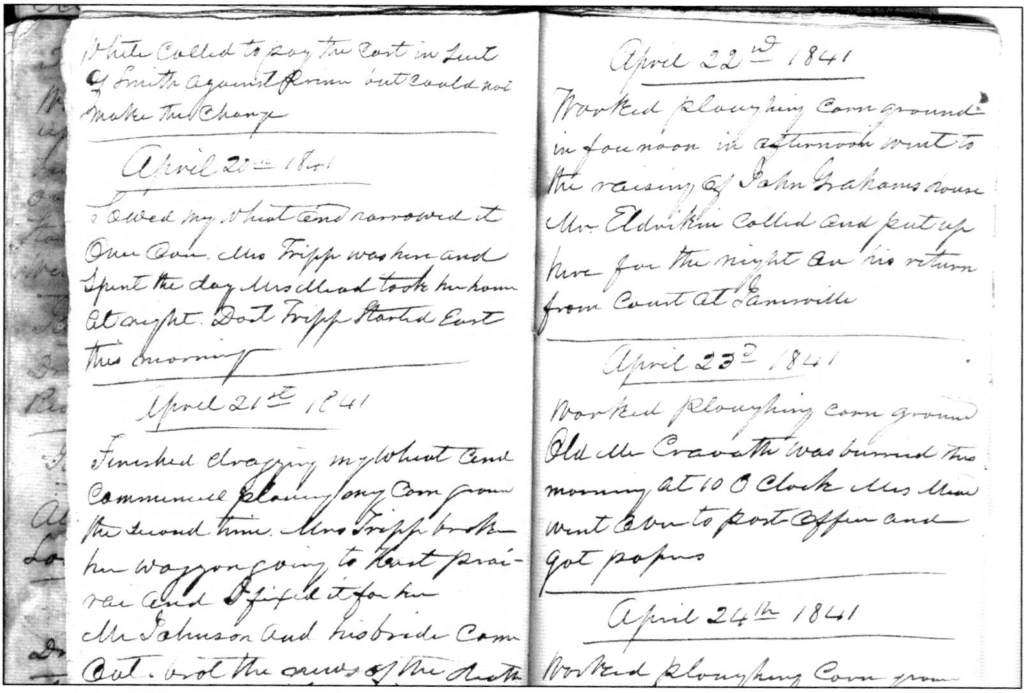

White Called to pay the Cost in Suit of Smith against Brown but could not make the Change

April 20th 1841

Sowed my Wheat and harrowed it Over Once. Mrs Tripp was here and Spent the day Mrs Mead took her home at night. Doct Tripp Started East this morning

April 21st 1841

Finished dragging my Wheat and Commenced ploughing my Corn ground the second time. Mrs Tripp broke her waggon going to Heart prairie and I fixed it for her Mr Johnson and his bride Came Out. brot the news of the death

April 22d 1841

Worked ploughing Corn ground in forenoon in afternoon went to the raising of John Grahams house Mr Eldridge Called and put up here for the night on his return from Court at Janesville

April 23d 1841

Worked ploughing Corn ground Old Mr Cravath was buried this morning at 10 O Clock Mrs Mead went down to post Office and got papers

April 24th 1841

Worked ploughing Corn ground

Zerah Mead, Esq., came to Whitewater with his family in May 1837, and he quickly made a name for himself with his "mammoth" 18-by-22-foot home, whose "crowning glory was a shingle roof." Gov. Henry Dodge appointed Mead justice of the peace for Walworth County in 1838. Mead was elected road commissioner for Whitewater in 1842 and elected to the state assembly in 1852. His diary is one of the few surviving written accounts of everyday life in early Whitewater. Pictured is Mead's entry from April 23, 1841, describing the burial of Deacon Prosper Cravath, who had died the day before. (WHS.)

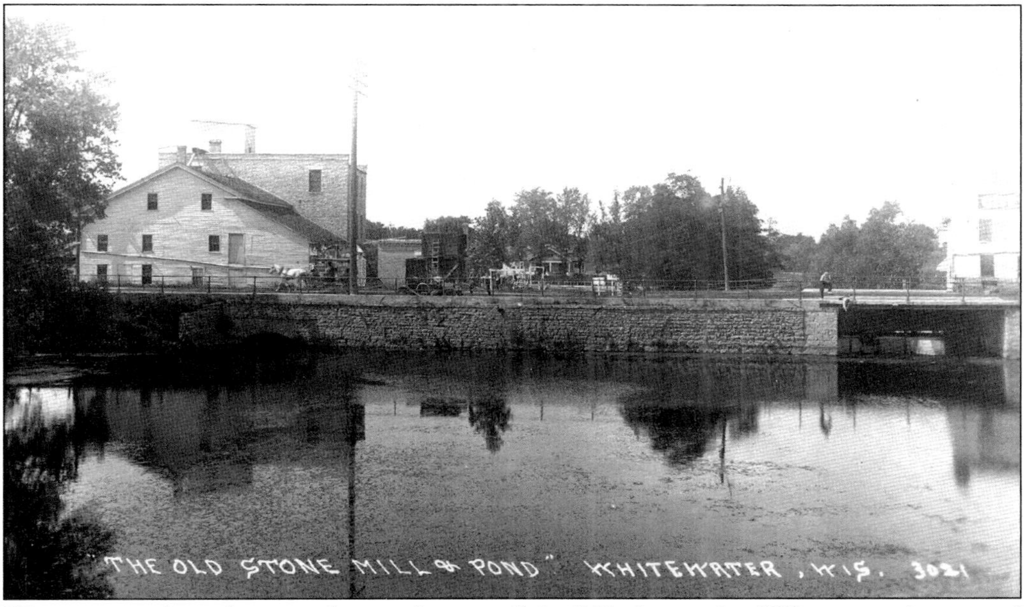

Whitewater industry began with a simple gristmill. In 1838, the people of Whitewater determined there was a need for a mill but that there was only one suitable mill site in town, on land owned by Daniel Butts. Butts relinquished his claim, and Dr. James Trippe provided the bonds necessary to build the mill. Construction began on June 27, 1839. (Irvin L. Young Memorial Library.)

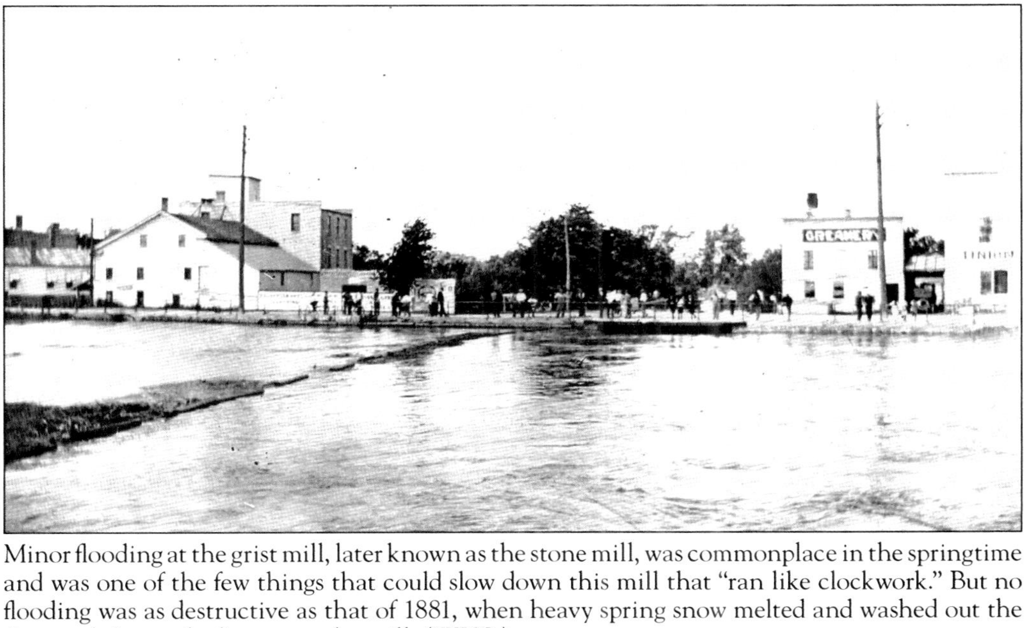

Minor flooding at the grist mill, later known as the stone mill, was commonplace in the springtime and was one of the few things that could slow down this mill that "ran like clockwork." But no flooding was as destructive as that of 1881, when heavy spring snow melted and washed out the dam and the road adjacent to the mill. (WHS.)

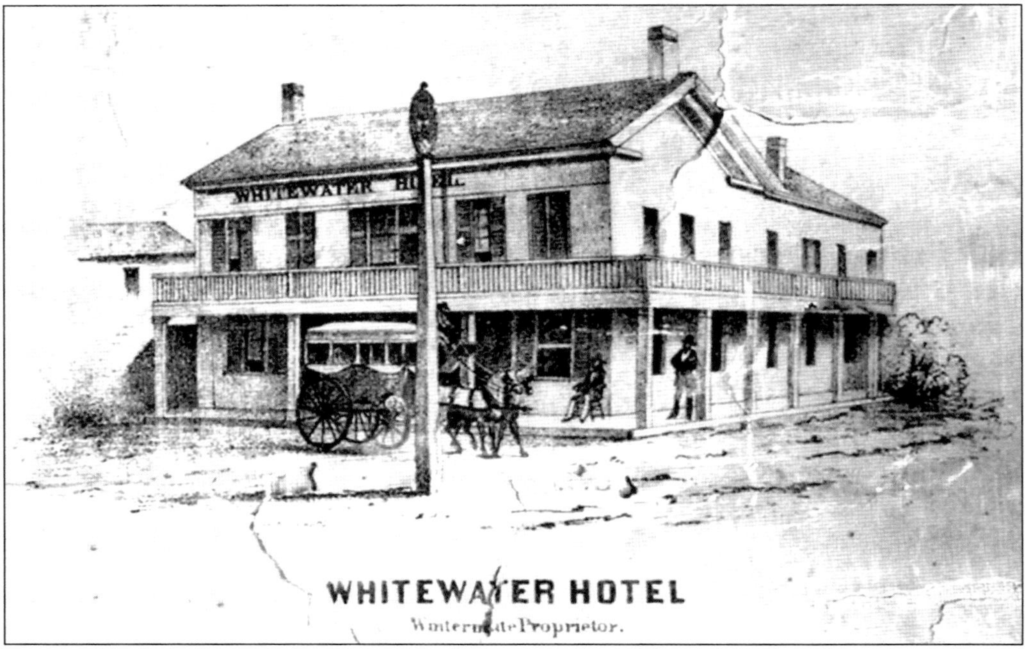

The Powers House was the first hotel built in Whitewater in 1840. It was constructed by David J. Powers on land owned and donated by Dr. James Trippe. It became known as the Whitewater Hotel the next year when it was purchased by Freeman Pratt. In 1858, Septer Wintermute razed the hotel and constructed the Montour House, another hotel, on the site. This photograph shows the original hotel in 1857. (University of Wisconsin–Whitewater, Andersen Library.)

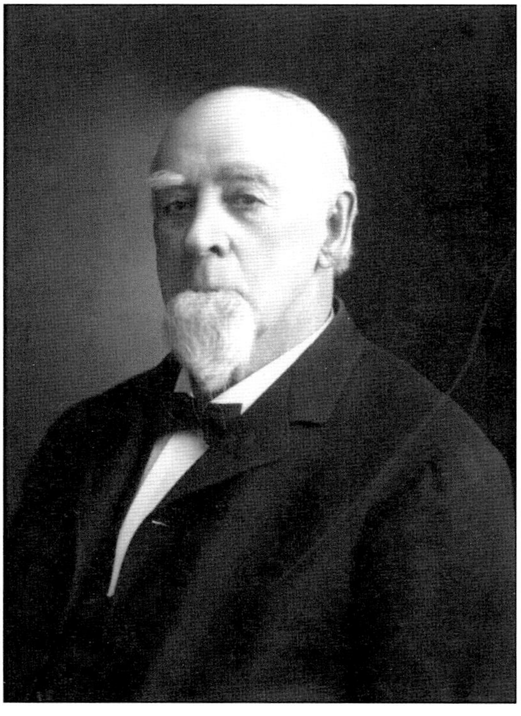

After arriving in Milwaukee in 1838, David J. Powers came to Whitewater. He eventually built his house on the corner of Fourth Street and Main Street. Powers was appointed postmaster by Martin Van Buren in 1838 and traveled on foot to Troy to be sworn in. He returned the same day, on foot, carrying Whitewater's first mail. (WHS.)

Philander Peck came to Whitewater during the summer of 1841. He hired David J. Powers to build him a store that measured just 28 feet by 18 feet. The store was located on the corner of Main Street and North First Street. Peck was a wealthy man who offered Whitewater's citizens long credits when purchasing goods, and other local businesses were eventually forced to offer similar credit plans. (Irvin L. Young Memorial Library.)

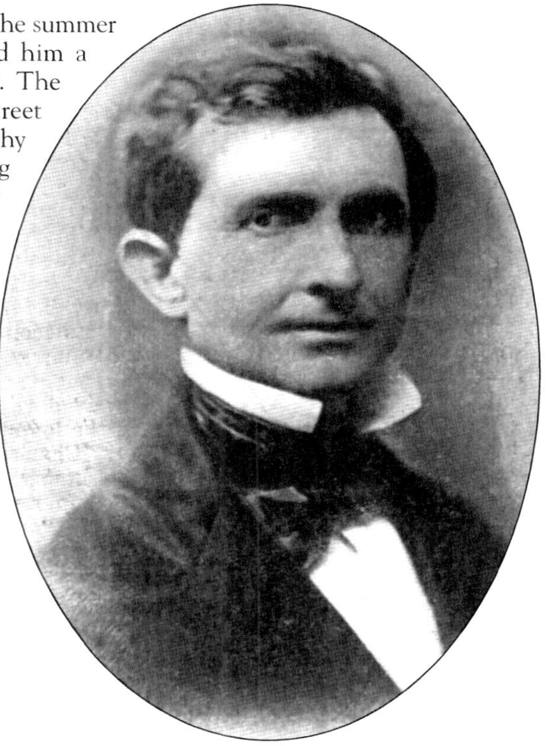

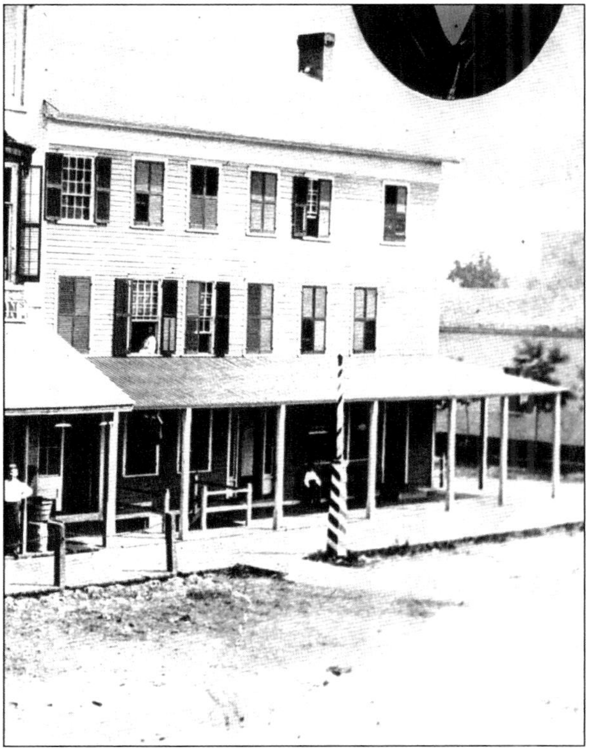

The Whitewater Exchange Hotel was built by Nehemiah Parsons and Jedediah Brown in 1842. C. E. Curtice took over and remodeled the hotel in 1857. The hotel was located on the northwest corner of Main Street and First Street. "It was opened on the total abstinence principle and ever to be a temperance house," according to Prosper Cravath, one of Whitewater's first settlers. (WHS.)

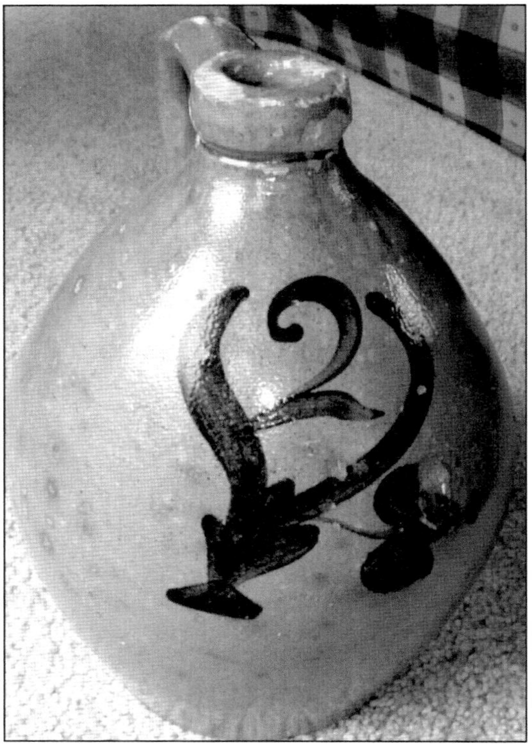

Daniel Cole first produced pottery in 1845 on the northwest corner of North Street and Fremont Street. Cole took on partner William Hunter, and the business moved to James Street, directly south of the railroad depot. They specialized in the production of flowerpots. Whitewater pottery is most often decorated with cobalt, which turns purple-black on red clay. Most pots were identified with gallon numbers but were not identified by maker. (MrBottles.com.)

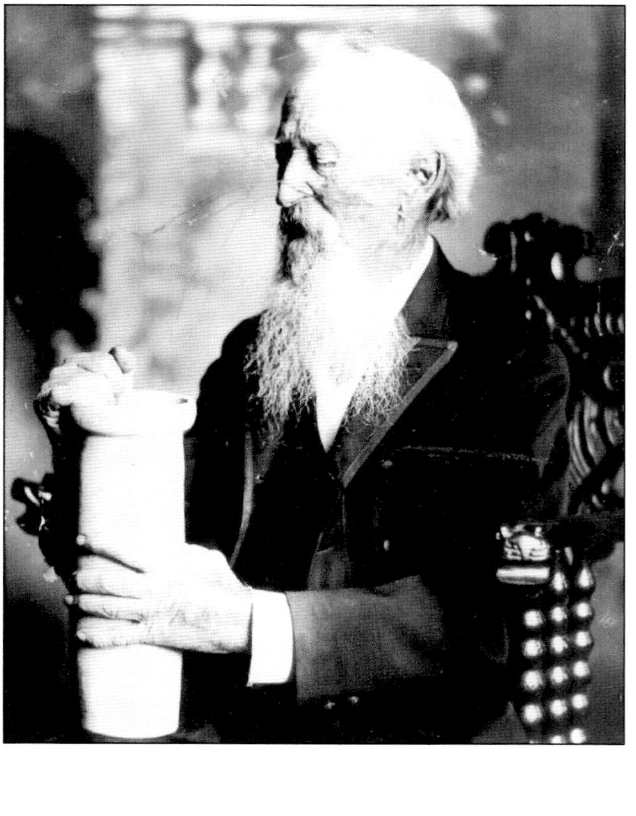

J. C. Williams and Company produced pottery from 1855 to 1859, after which Dan Cole and Company assumed the business until 1871. Another shop in Whitewater was run by Michael Ohnhaus and John Milz. The business was located on Summit Street and produced pottery from 1859 to 1881. Pictured is Michael Ohnhaus holding a drainage tile he had produced years before. (Maybelle Ohnhaus, Kori Oberle.)

The potteries used local clay and lead glaze to produce earthenware, also referred to as yellowware or redware. The earthenware had a distinctive yellowish-orange appearance. Cole and Hunter used more than 15,000 pounds of lead a year to create their glazes. The glaze on this cream pot is yellow with orange specks and is marked with the number 4 to designate its capacity in gallons. (MrBottles.com.)

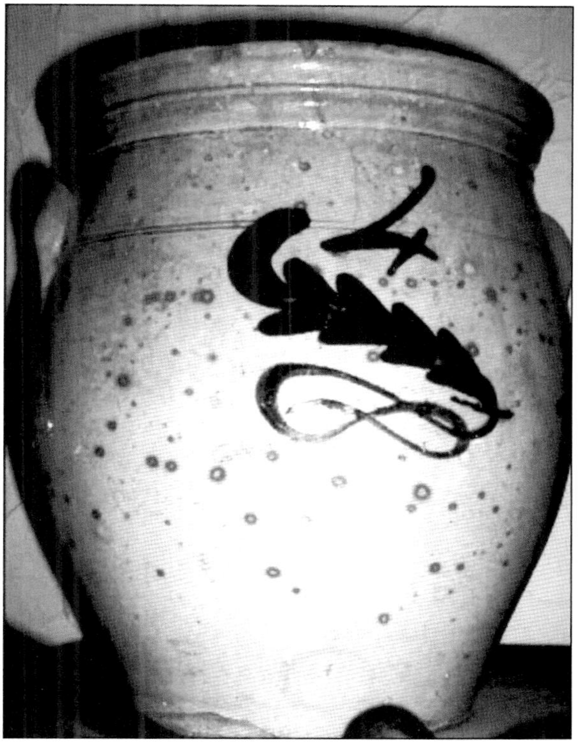

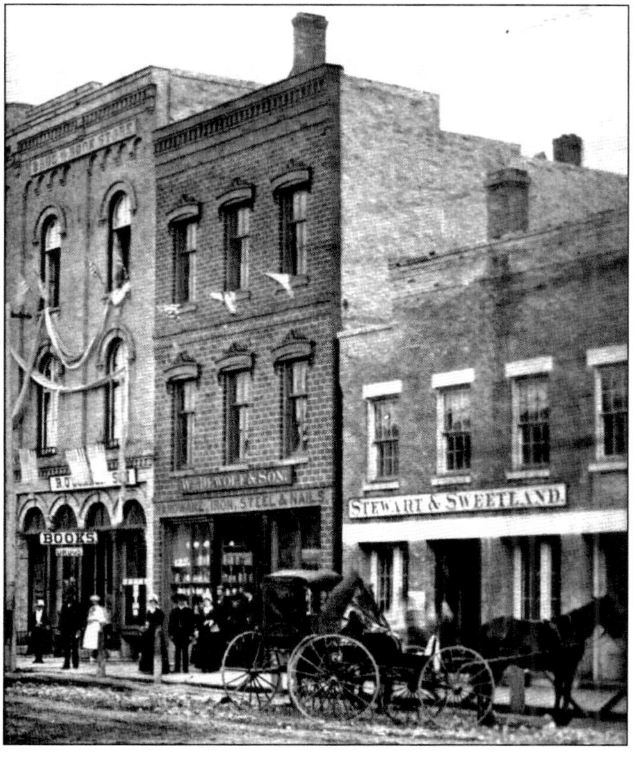

The shop of Cushing, Sweetland, and Stewart dealt in dry goods, groceries, hardware, iron, steel, nails, paints, oils, and cooper's tools, as well as boots, shoes, and other leather goods. Joseph Cushing, John D. Sweetland, and William Stewart were skilled self-promoters who promised that their goods were sold at prices "as low, if not lower than the same quality of goods can be, anywhere else west of Milwaukee." (Irvin L. Young Memorial Library; Fran Achen Collection.)

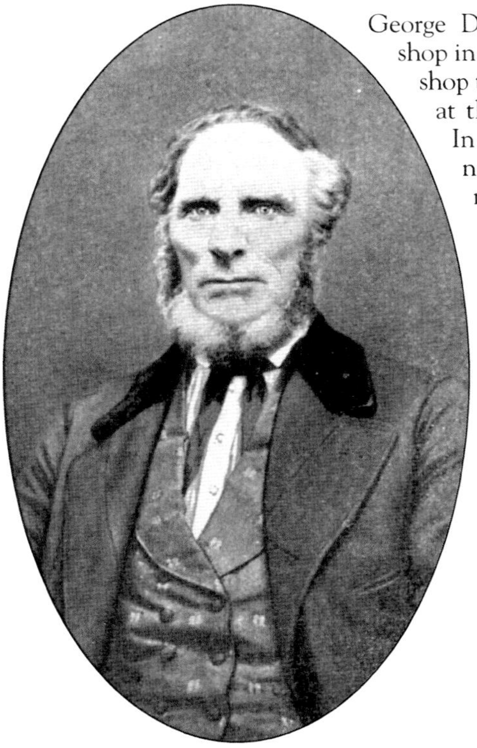

George Dann, pictured, started Whitewater's first cooper shop in 1845, located on Fourth Street. He then moved his shop to the east side of Whitewater Creek two years later at the corner of Jefferson Street and Cravath Street. In 1847, Dann built Whitewater's second brickyard near Cravath Street and Jefferson Street, producing more than a million cream-colored bricks by 1861. (WHS.)

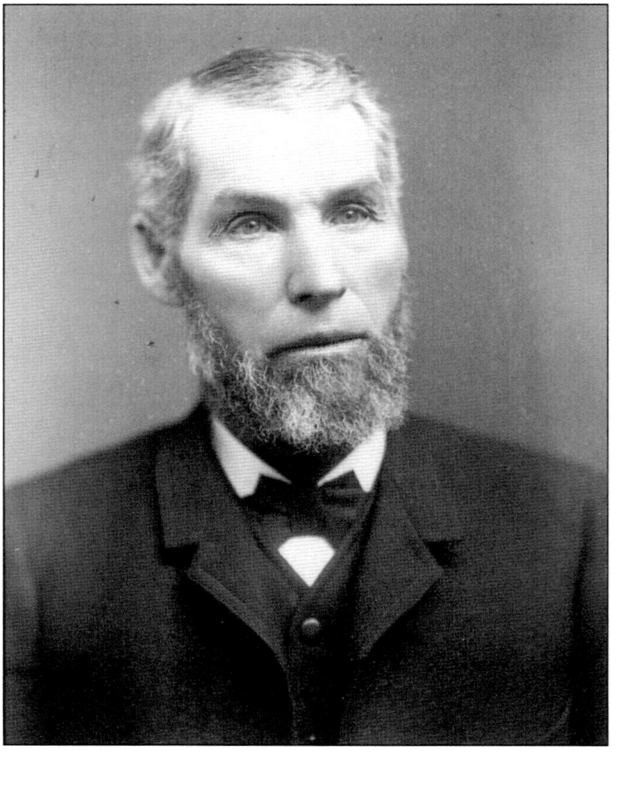

In 1879, Joseph Dann (pictured) and Edward Roethe opened a brickyard behind the hill north of Cravath Street. Joseph Dann, a former forty-niner of the California Gold Rush, manufactured farm and market baskets before becoming a brickmaker in 1873. Dann sold the business to Charles Martin in 1891. It closed during the panic of 1893. (WHS.)

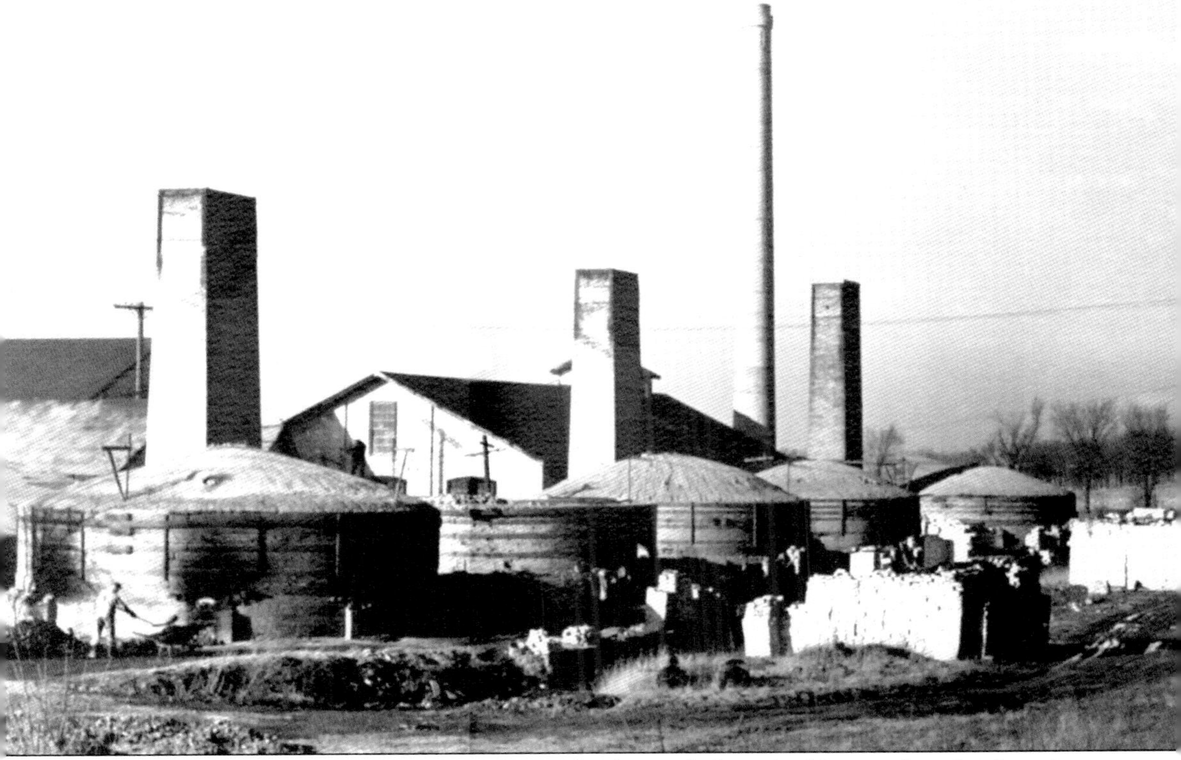

The supply of Whitewater's bricks very rarely met the demand of area builders, as these brickyards helped accelerate Whitewater's growth. While the two brickyards owned by the Danns were the most significant, many smaller yards made their home in Whitewater. The first brickyard in Whitewater was owned by William Wood and began production on May 1, 1841. It was located on the south side of the railroad tracks near Cravath Lake. Albert Kendall opened a brickyard on the west side of Fremont Street in 1852 that operated until the 1870s. Whitewater Brick and Tile, pictured here, was built in 1903 and closed during the 1940s. The bricks created here turned a signature cream color because of the action of heat on lime, calcium, and magnesium found in Whitewater's natural clay deposits. (Irvin L. Young Memorial Library; Fran Achen Collection.)

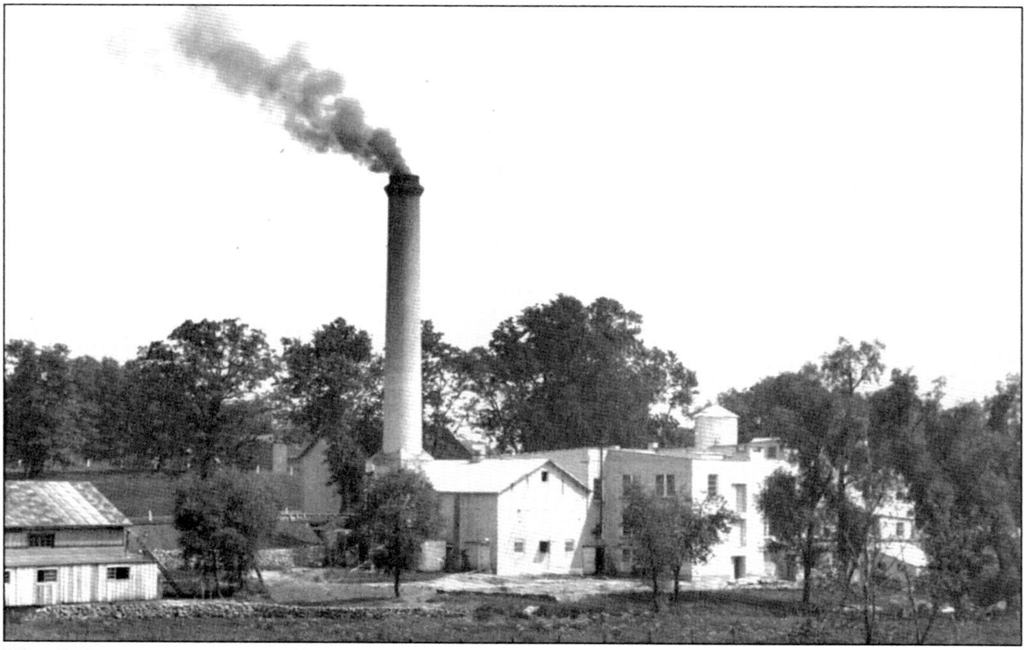

The Whitewater Paper Mills were built in 1860 by Daniel Tripp and John M. Crombie. They initially produced newspaper and book paper. In 1866, the mills were sold to John W. Denison and Leonard A. Tanner. They were destroyed by fire in 1868 and rebuilt the next year. The new two-story mill building turned out 3,000–4,000 pounds of brown wrapping paper each day. (Irvin L. Young Memorial Library; Fran Achen Collection.)

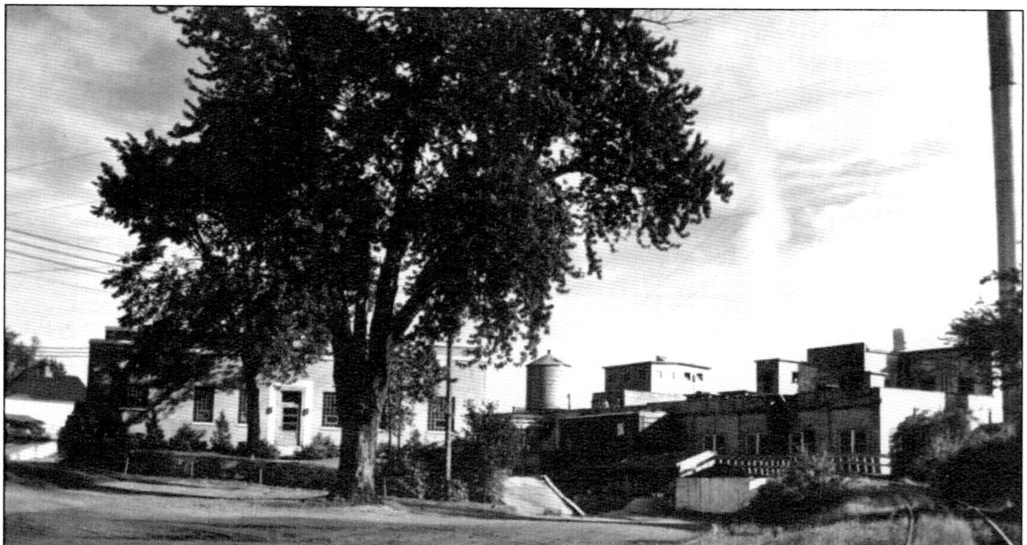

In the 1870s, the Whitewater Paper Mill produced straw wrapping paper, white and colored tea papers, and print paper. Charles Allen and Benjamin Frees were the last owners of the mill before it was sold to the Columbia Straw-Board Company in 1894. The mill closed shortly thereafter because of a scarcity of straw and a decreased demand for straw paper. The mill site later became the site of the condensery. (Irvin L. Young Memorial Library; Fran Achen Collection.)

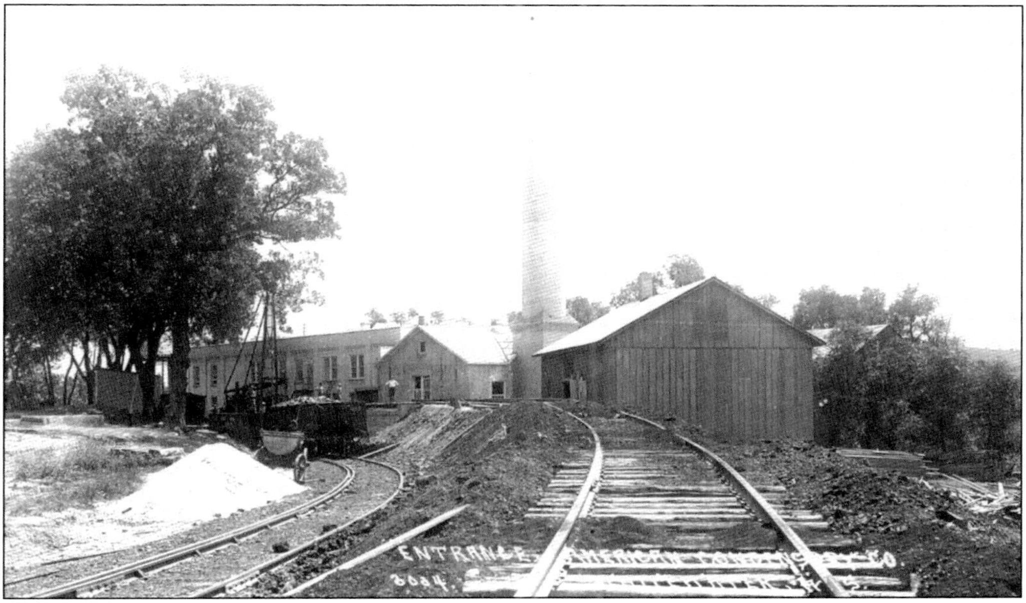

Most early creameries were built with an adjacent icehouse. Icehouses like the one pictured here allowed trains to pick up ice to use in their milk cars. The daily milk train would arrive to pick up dairy products before proceeding to drop off passengers and mail at the depot. In 1873, legendary Wisconsin dairyman W. D. Hoard obtained the first refrigerator car, allowing the shipment of Wisconsin cheese to eastern markets. (Irvin L. Young Memorial Library; Fran Achen Collection.)

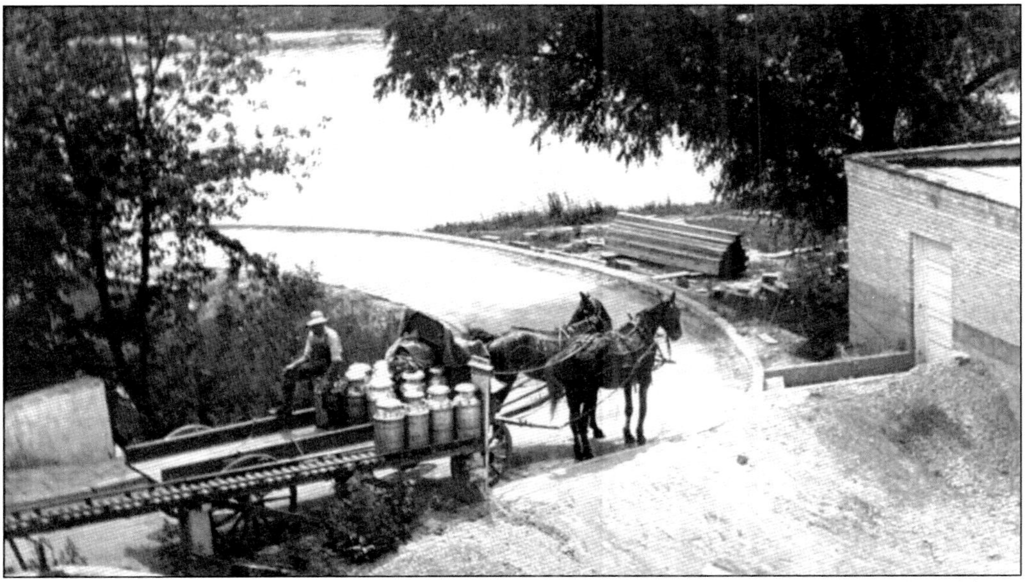

Area farmers had no trouble finding a market for their milk, as the condensery produced large quantities of dairy products in a central location. Daily deliveries from farmers ensured the milk was fresh. Before the milk bottle was in common use, dairies distributed milk in wagons equipped with large cans or crocks, and the milk was ladled into the customer's bucket from the wagon. (Irvin L. Young Memorial Library; Fran Achen Collection.)

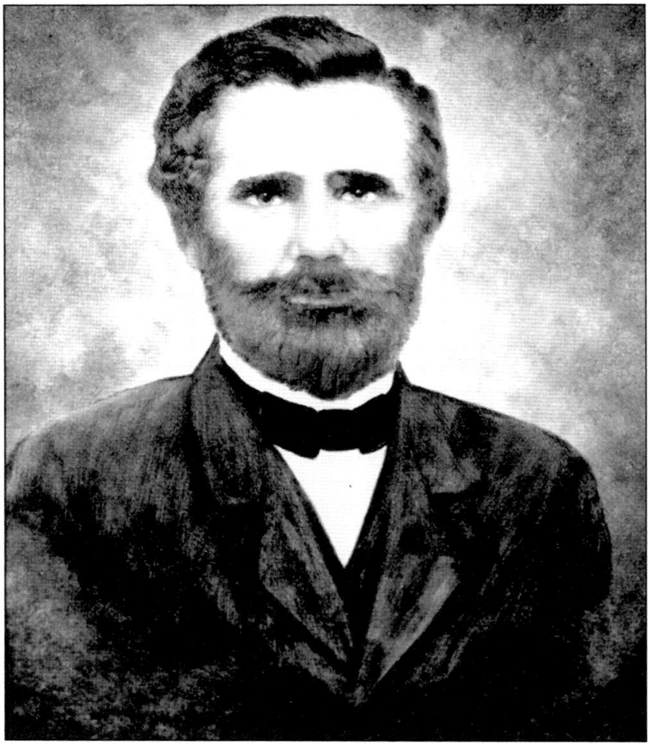

Sanger Marsh came to Whitewater in 1845 and went into business with John S. Partridge, opening a general store. He worked in the business until 1855. In 1863, Marsh became the first president of the First National Bank of Whitewater, only the second national bank in Wisconsin. He died on October 29, 1872, when his horse bolted and overturned his carriage. (First Citizens State Bank.)

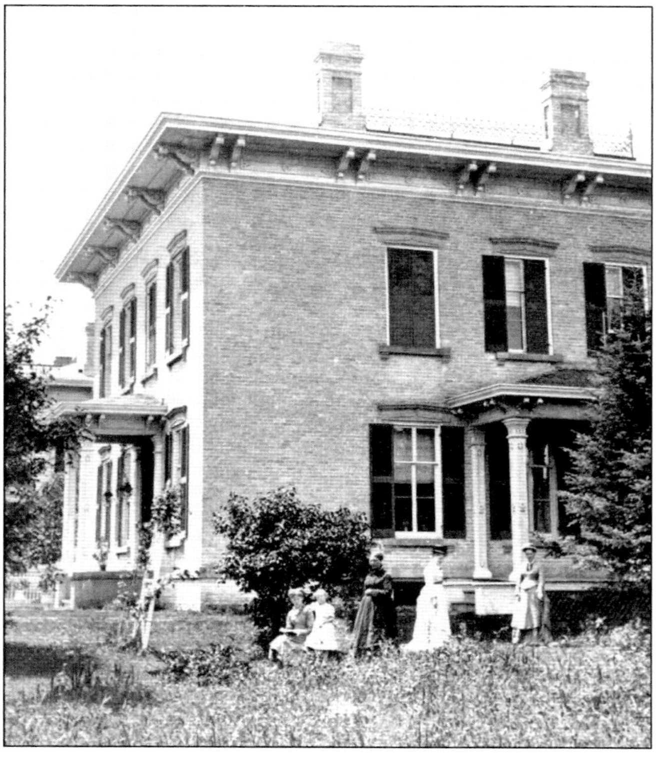

The Sanger Marsh Home on Main Street was built in 1861 by Nelson Salisbury, who built many of the large, fine homes in Whitewater. Marsh had three young daughters by this time with his second wife, Chelsea Pratt. The house was originally situated on a plot of more than nine acres that reached back to Starin Road. (WHS.)

Pictured is Sanger Marsh's son, George Sanger Marsh, in 1918. He is shown with his second wife, the former Rebecca Jane Wight, whom he married in 1874. In 1872, George Sanger Marsh became a cashier at his father's business, the First National Bank. George later became president of the Citizens State Bank of Whitewater, incorporated in 1883. (WHS.)

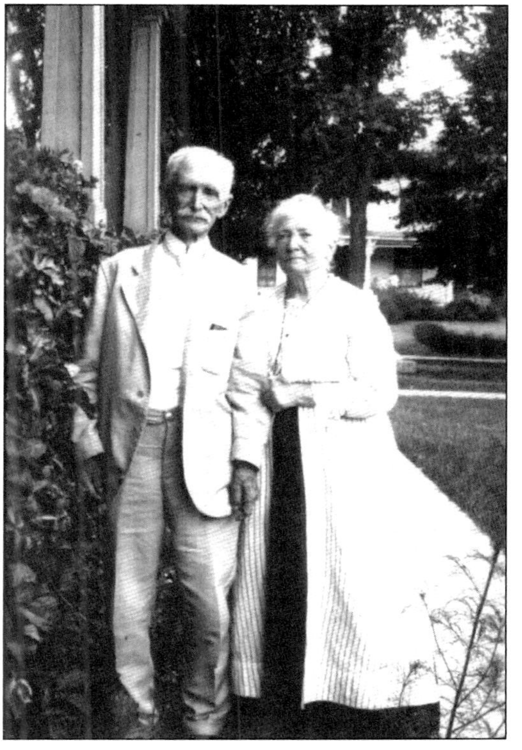

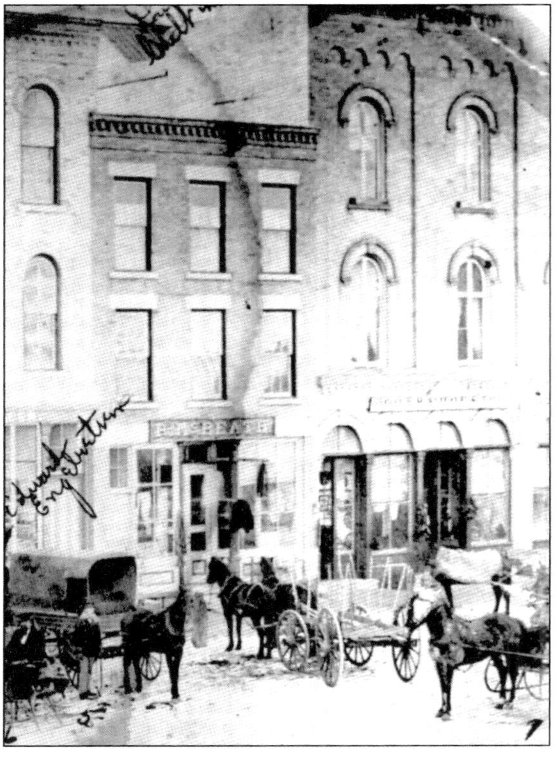

Robert McBeath was a Scottish-born tailor who established his business in 1850, in conjunction with Patrick McLaughlin. The partners built the first three-story building in Whitewater in 1854. When McLaughlin died in 1858, McBeath continued the business, eventually doing more than $10,000 of business annually. McBeath served on the town board of supervisors and the village board. He also worked for the American Bible Society. (Irvin L. Young Memorial Library; Fran Achen Collection.)

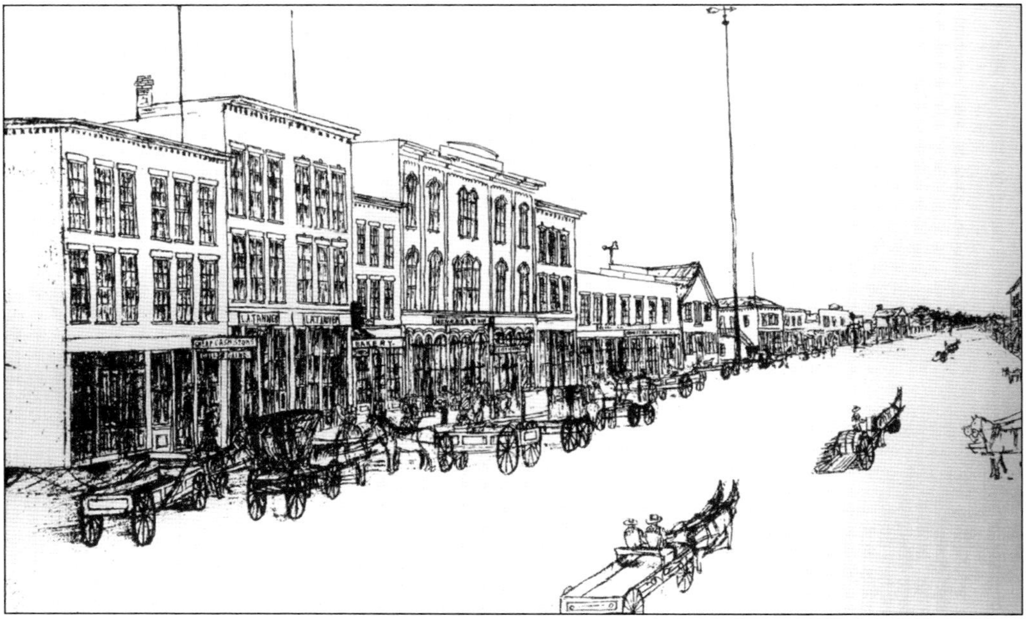

Orlando Gallt started his general merchandise business in 1854 in a small store on Main Street with his sons-in-law Edwin Caine and Warren Cole. The operators of Gallt and Caine's Cheap Cash Store were said to have had such a passion for money "they will sell anything they have, cheaper than any other store in town, to obtain it." This illustration was drawn by employee Henry Rile. (WHS.)

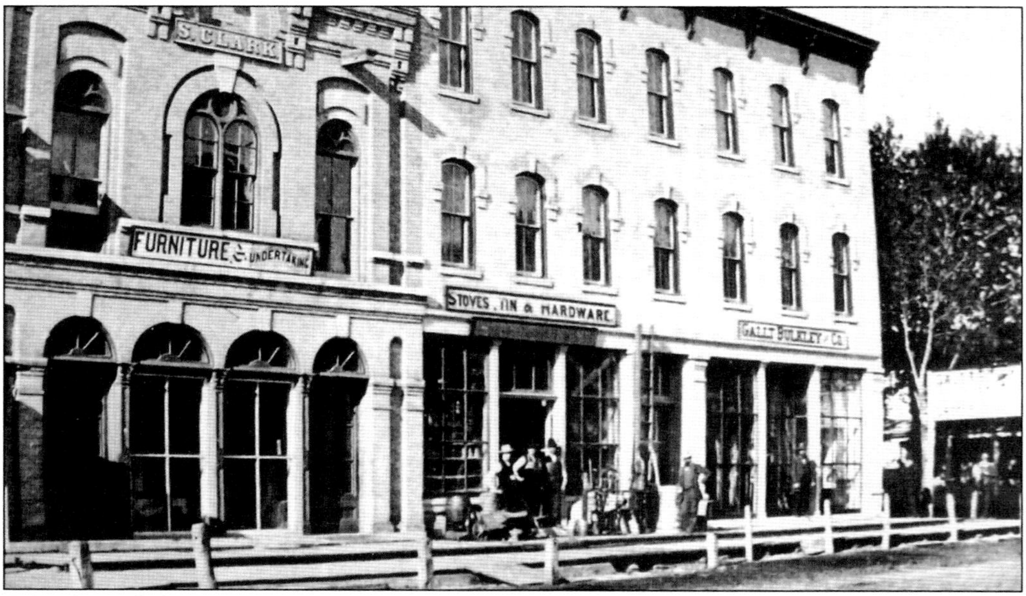

Orlando Gallt later opened a store as part of Gallt, Bulkley and Company, located on the corner of Second Street and Main Street. After Gallt retired in 1882, the new Bulkley, Crandall and Company dealt in dry goods, carpets, groceries, and crockery. Brownell Bulkley formed this new partnership with his brother-in-law Elias Crandall, and Gallt's daughter Helen Caine remained to represent the Gallt family's interests. (Irvin L. Young Memorial Library; Fran Achen Collection.)

Edwin Caine of Gallt and Caine died in 1868, leaving Helen Mary Gallt Caine as his widow. Edwin left her with two small children, Edwin and Anna. She lived with her parents after Edwin's death, working in her father's business as a saleswoman and family representative. She died in 1907 at her daughter's home in Nebraska. (Stephanie Ostler.)

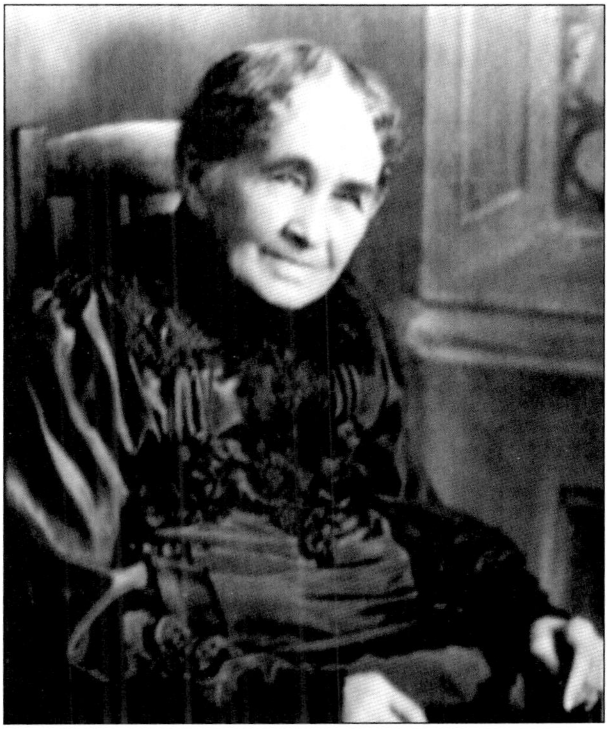

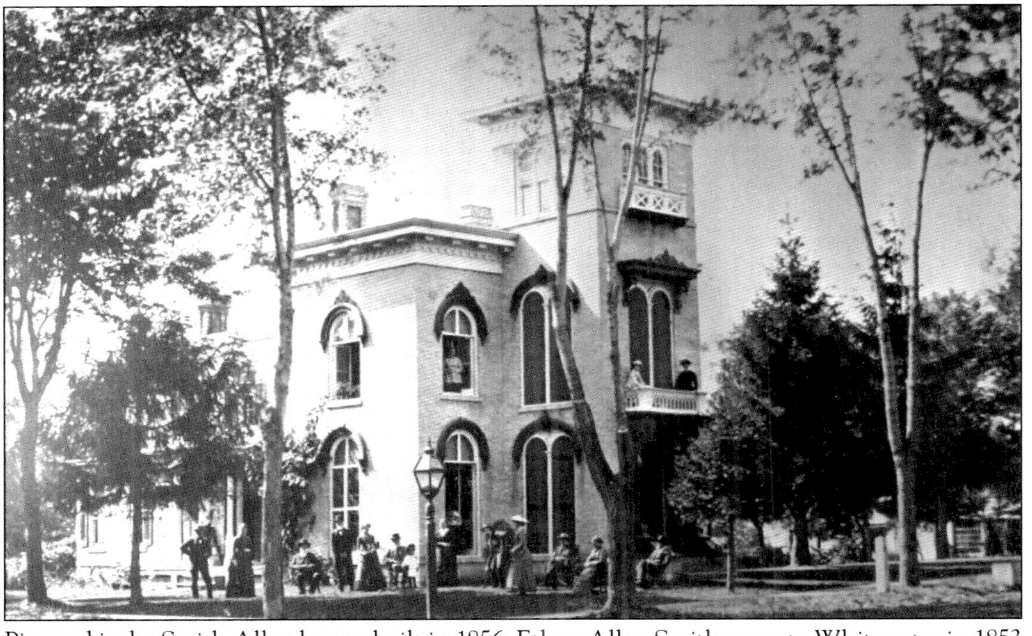

Pictured is the Smith-Allen house, built in 1856. Ethan Allen Smith came to Whitewater in 1853 and set up a store in the "Emporium Block," the first building on the south side of Main Street. He operated a successful dry goods shop until 1857, when he left Wisconsin and moved to Kansas. Nathan Allen, a lumberman, then acquired the house. (Irvin L. Young Memorial Library; Fran Achen Collection.)

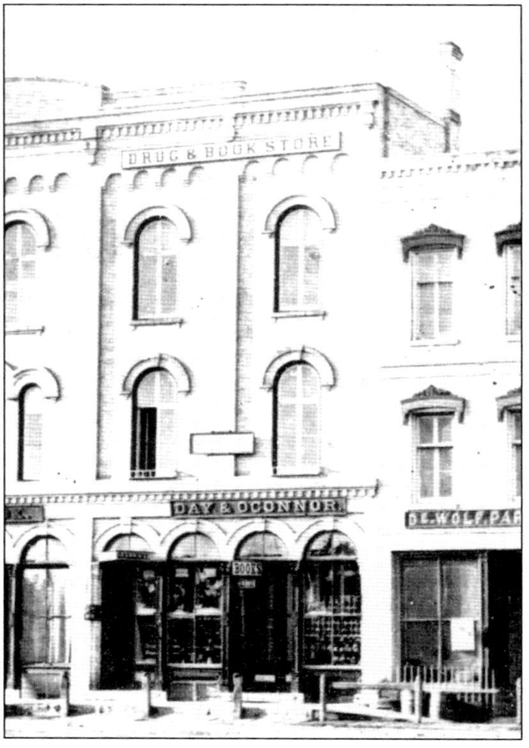

Richard J. O'Connor, druggist and apothecary, came to Whitewater in 1842. In 1846, he founded his own apothecary shop before opening the Day and O'Connor apothecary shop (pictured) with Ira Church Day starting in 1862. Early apothecaries were pharmacists as well as general doctors, surgeons, and midwives. Richard also served as town assessor for more than 20 years. His sons Fred and Harry continued the business into the 1920s. (Irvin L. Young Memorial Library; Fran Achen Collection.)

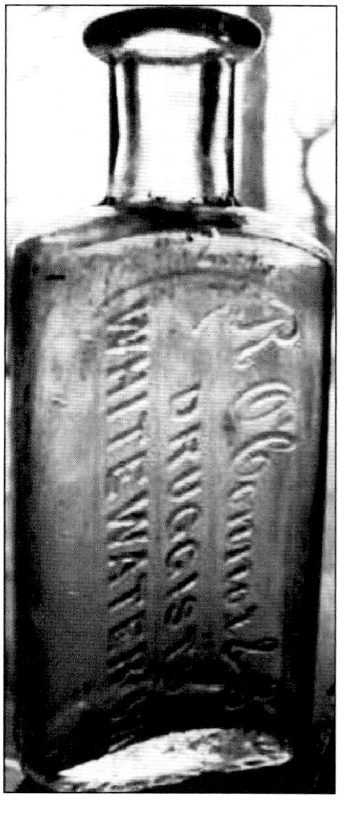

Nineteenth-century apothecaries dispensed their own prescriptions. Day and O'Connor offered "the best drugs and chemicals in use," and they carried their own line of personalized containers and bottles. This four-and-one-quarter-inch bottle is from O'Connor's original store. The base is marked with "DF&CO," denoting that the bottle was made by Chicago's Dean, Foster and Company. (MrBottles.com.)

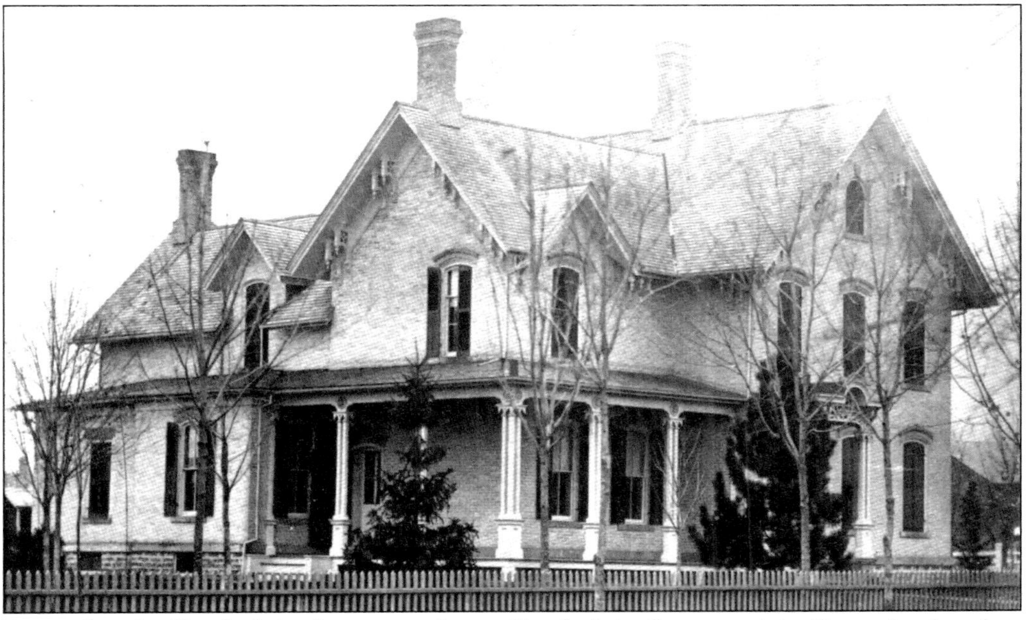

Pictured is the Hanford A. Conger residence. Hanford A. Conger, originally a schoolteacher, opened a horticultural nursery in Whitewater in 1854. Hanford Conger's son Eugene joined him in the business, which specialized in fruit trees and shade trees. Later Conger added 200 acres to his holdings, just east of Whitewater. The Conger home was located at the corner of Prairie Street and Main Street. (WHS.)

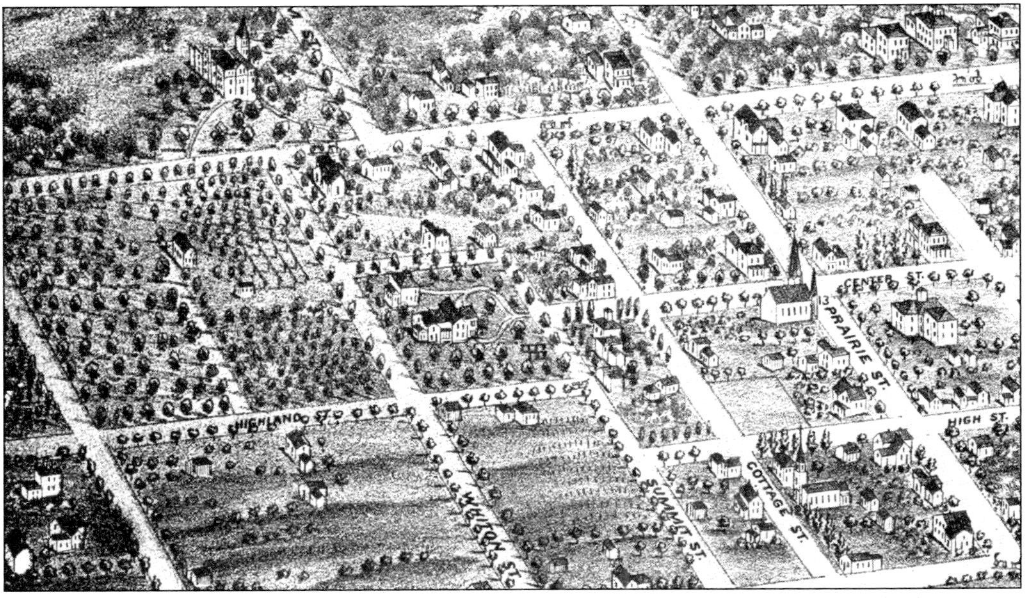

This section of an 1870 map of Whitewater illustrates the area occupied by Hanford Conger's nursery, left center. The nursery was bounded by Main Street, South Prince Street, Highland Street, and Whiton Street. The small building pictured in the center of the nursery was used as the nursery office. This building still stands on Conger Street, which now intersects the former site of the nursery. (WHS.)

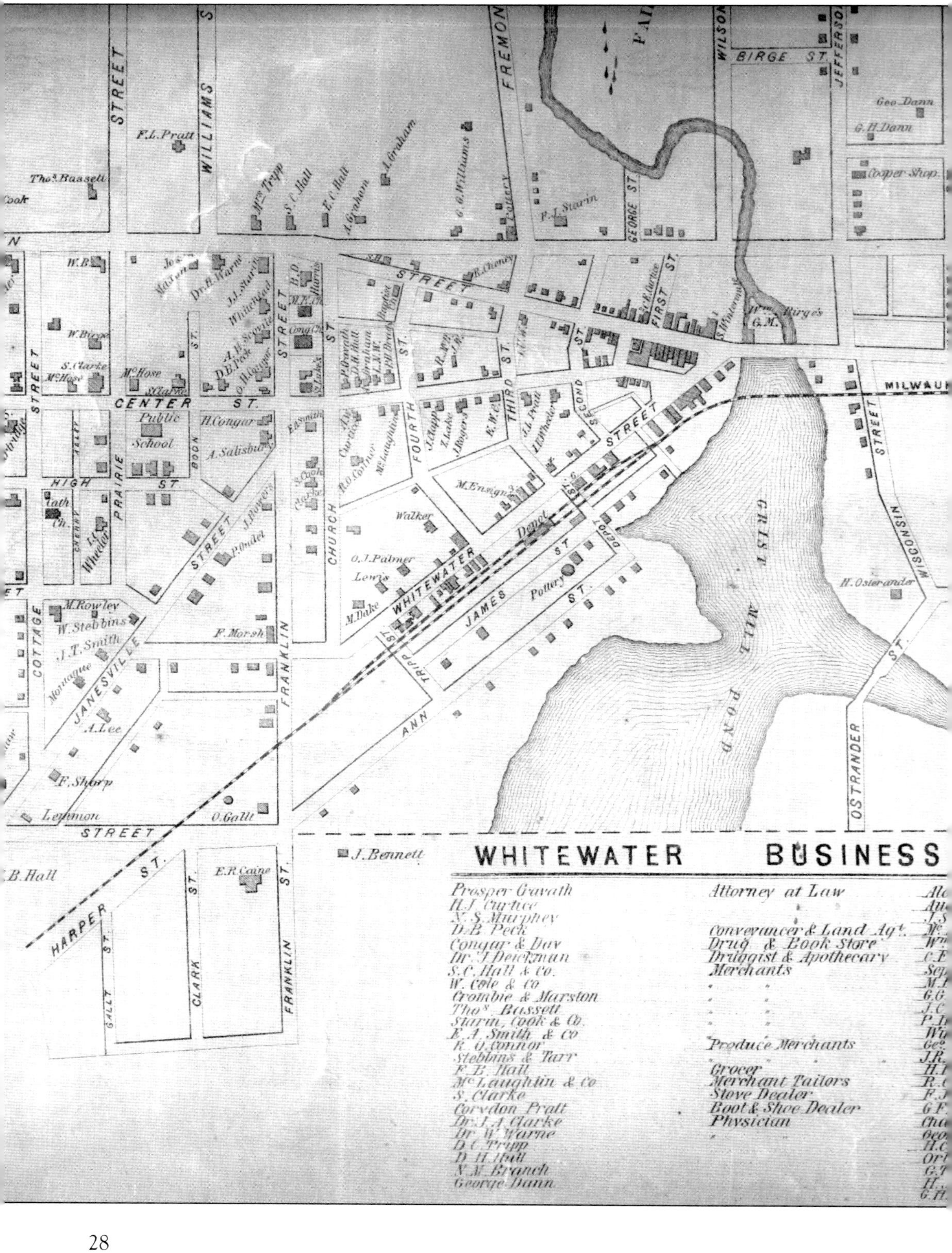
WHITEWATER BUSINESS
Prosper Gravath
H.J. Curtice
N.S. Murphey
D.B. Peck
Congar & Day
Dr. J. Deichman
S.C. Hall & Co.
W. Cole & Co
Crombie & Marston
Thos. Bassett
Starin, Cook & Co.
E.A. Smith & Co
R. O. Connor
Stebbins & Tarr
F.E. Hall
Mc Laughlin & Co
S. Clarke
Corydon Pratt
Dr. J.A. Clarke
Dr. W. Warne
D.C. Tripp
D.H. Hall
N.M. Branch
George Dann
Attorney at Law
Conveyancer & Land Agt.
Drug & Book Store
Druggist & Apothecary
Merchants
Produce Merchants
Grocer
Merchant Tailors
Stove Dealer
Boot & Shoe Dealer
Physician
STREET
WILLIAMS
FREMONT
WILSON
BIRGE ST.
JEFFERSON
GEORGE ST.
FIRST ST.
CENTER ST.
HIGH ST.
PRAIRIE
CHURCH
FOURTH ST.
THIRD ST.
SECOND ST.
WHITEWATER STREET
JAMES ST.
DEPOT ST.
TRIPP ST.
ANN
JANESVILLE STREET
COTTAGE
FRANKLIN
FRANKLIN ST.
HARPER ST.
GALLT ST.
CLARK ST.
MILWAUKEE
WISCONSIN STREET
OSTRANDER ST.
GRIST MILL POND
Public School
Depot
Pottery
Cooper Shop
Cath. Ch.
Cong. Ch.
M.E. Ch.
Baptist Ch.
F.L. Pratt
Thos. Bassett
S.C. Hall
E.C. Hall
A. Graham
J. Graham
G.G. Williams
Pottery
F.L. Starin
R. Cheney
Wm. Birge's G.M.
Geo. Dann
G.H. Dann
H. Congar
A. Salisbury
M. Ensign
Walker
O.J. Palmer
Lewis
M. Dake
M. Rowley
W. Stebbins
J.T. Smith
A. Lee
F. Sharp
Lemmon
O. Gallt
F. Morsh
J. Bennett
E.R. Caine
H. Osterander
E.B. Hall
P. Oudet
S. Clarke
W. Birge

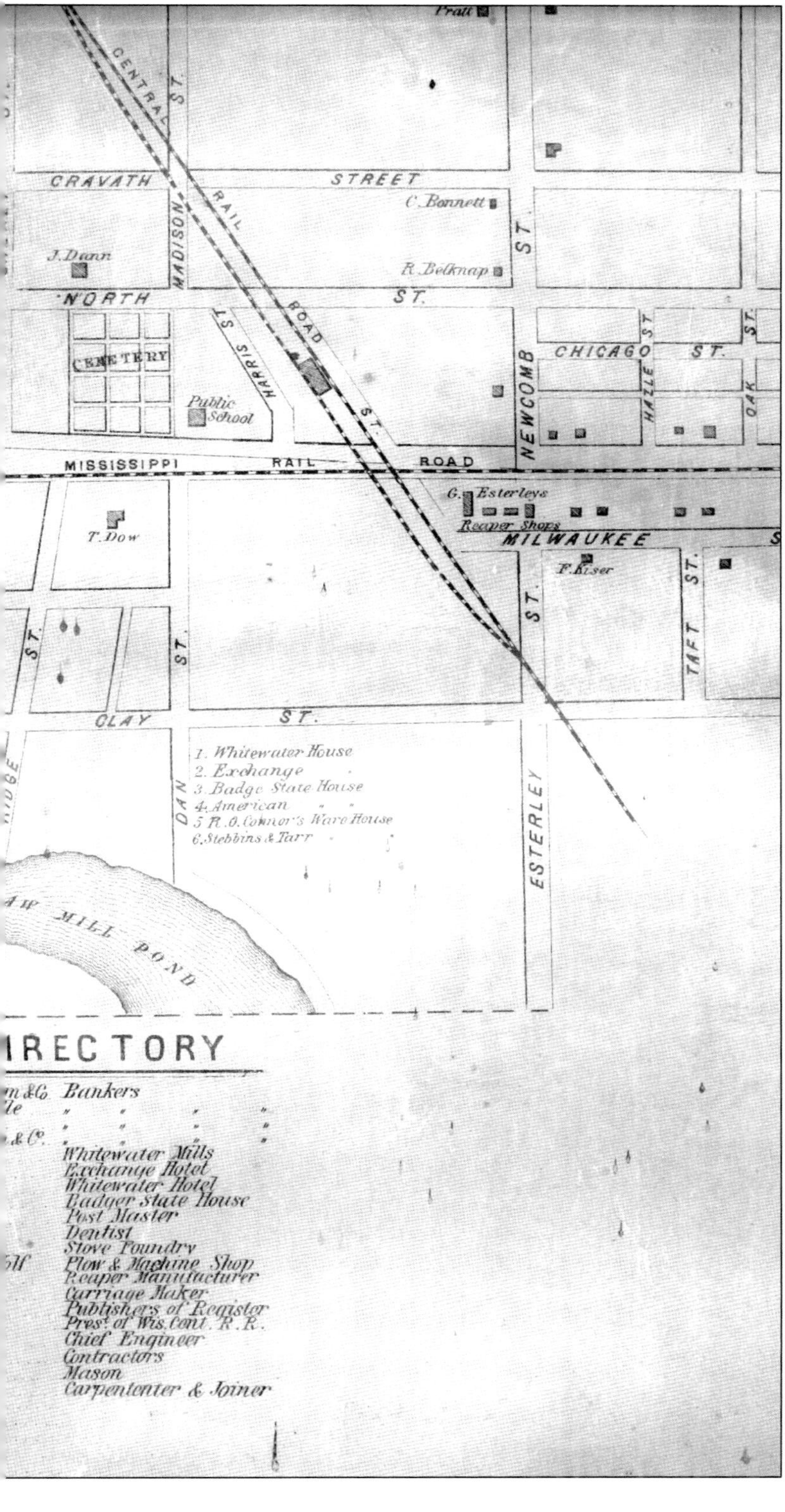

This 1857 plat map was published by T. R. Redding and J. J. Watson, city and county map publishers from Elkhorn. This was the earliest plat map published in Wisconsin. Redding and Watson included a brief description of Whitewater, characterizing the town as follows: "diversified with rolling prairies, shady groves, crystal lakes, winding streams and ever flowing streams of purest water . . . the people are generally intelligent, enterprising, and industrious . . . a healthier, more beautiful, or more productive country cannot be found." George Esterly's factory lies on the far right of the map, at the intersection of the Milwaukee and Mississippi Railroad and the proposed line of the Wisconsin Central railroad. The Wisconsin Central would never reach Whitewater. (University of Wisconsin–Whitewater, Andersen Library.)

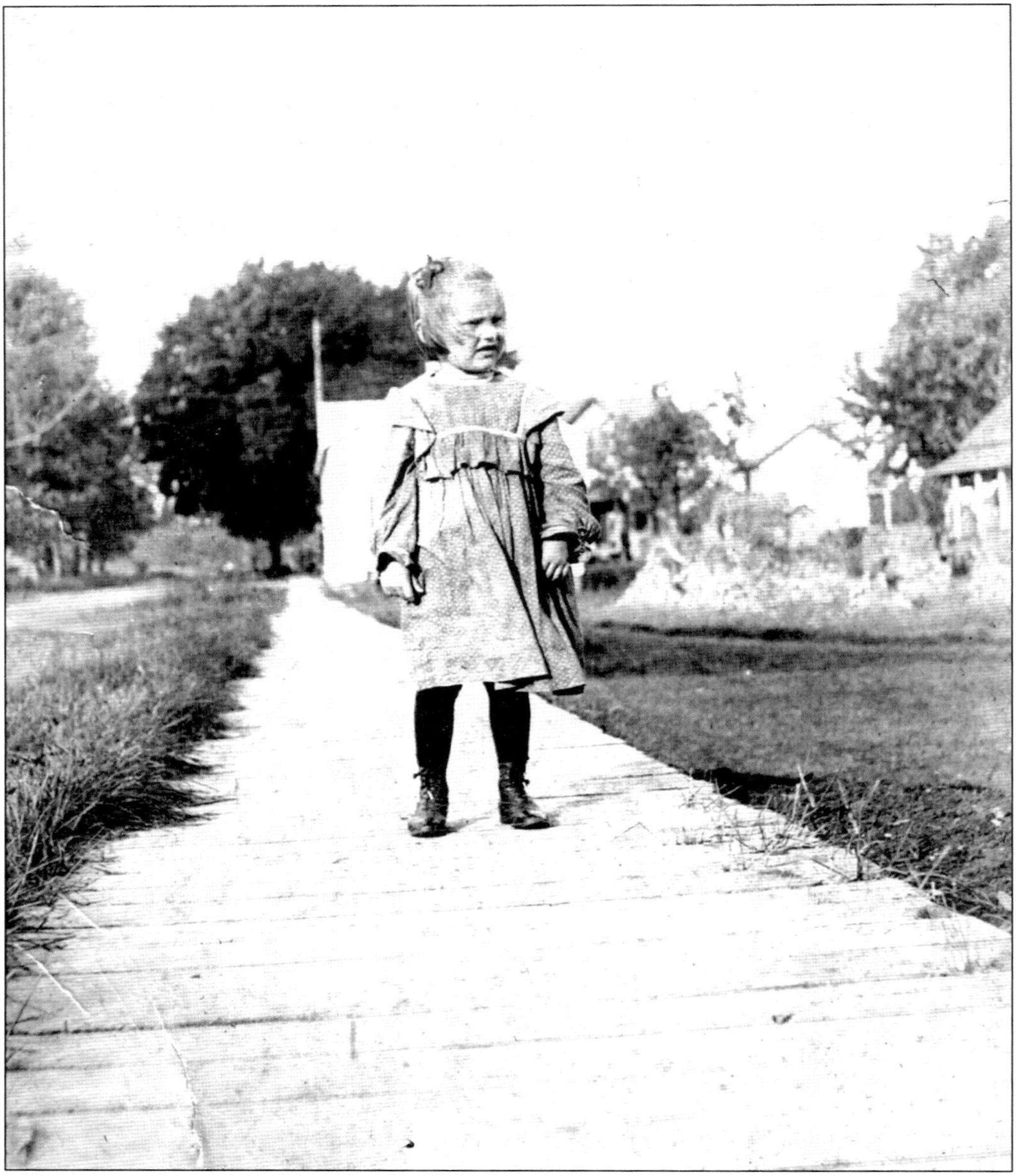

Early in 1858, a group of "spirited men gathered one morning, contrary to law, and laid the first cross-walk in front of the public school building on Center Street." After this act, which was met with great celebration, it was never again against the law to lay a crosswalk or sidewalk. However, early sidewalk construction was slow. By the end of July, the *Whitewater Register* decried the lack of approval for a Main Street sidewalk, saying that "there is always, in every town, some one or more citizens who are so uncompromisingly old fogy or so supremely selfish and miserly that they would not spend ten cents to keep the earth making its diurnal revolutions." This sidewalk was finally approved in August 1858. Pictured is young Ethel Henderson on a similar sidewalk around 1902. (Henderson family.)

Newton M. Littlejohn came to Whitewater in 1856 and became the town's first president after the village's incorporation in 1858. Littlejohn was a lumber dealer who sold more than three million feet of lumber a year. In 1863, he was elected to the state senate and served for six years as "an able and patriotic legislator." (First Citizens State Bank.)

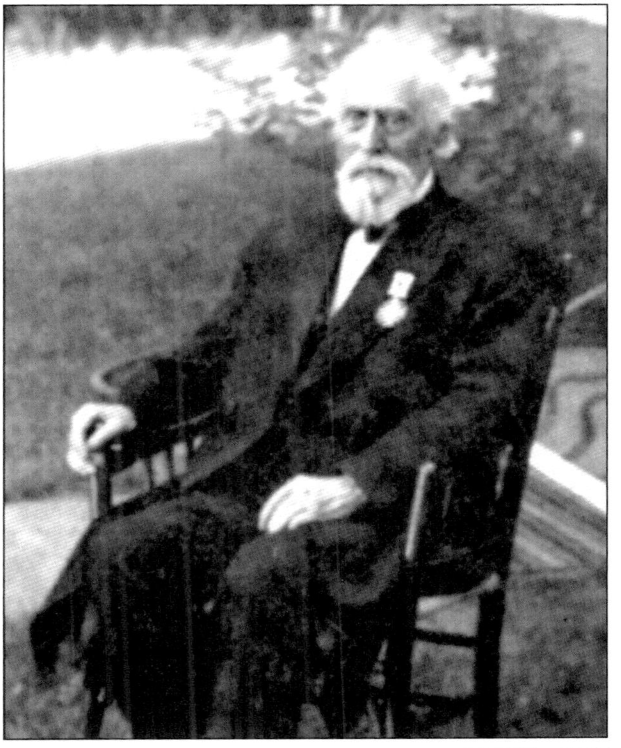

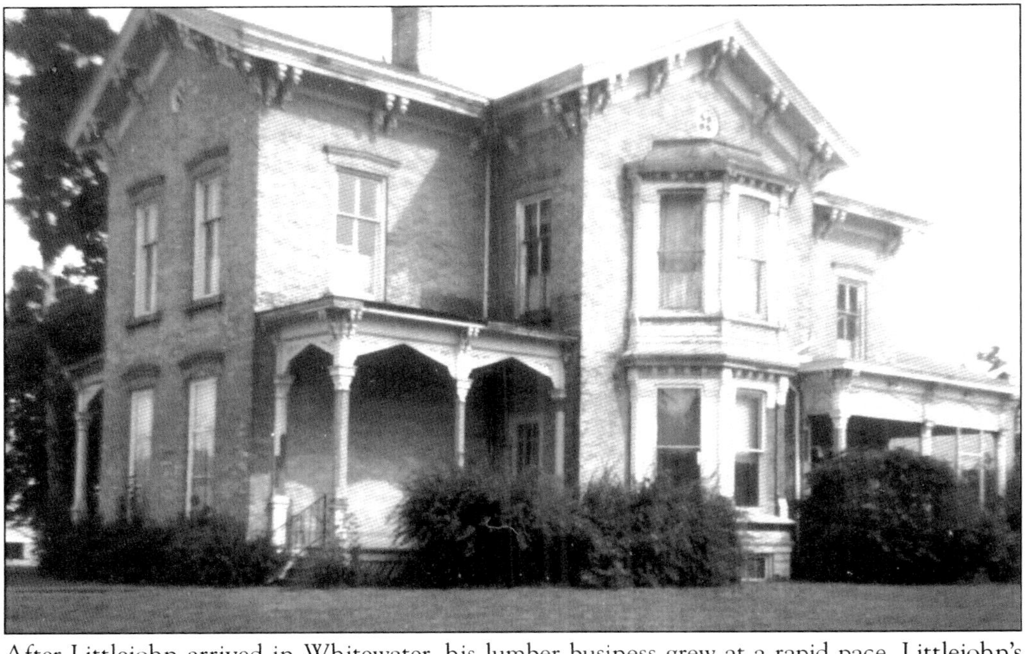

After Littlejohn arrived in Whitewater, his lumber business grew at a rapid pace. Littlejohn's early success allowed him to construct this huge brick house by 1859. A loyal Whitewater man, he remained in the house until his death in 1916. At one time, the Littlejohn residence was used as a parish house for the First English Lutheran Church. (WHS.)

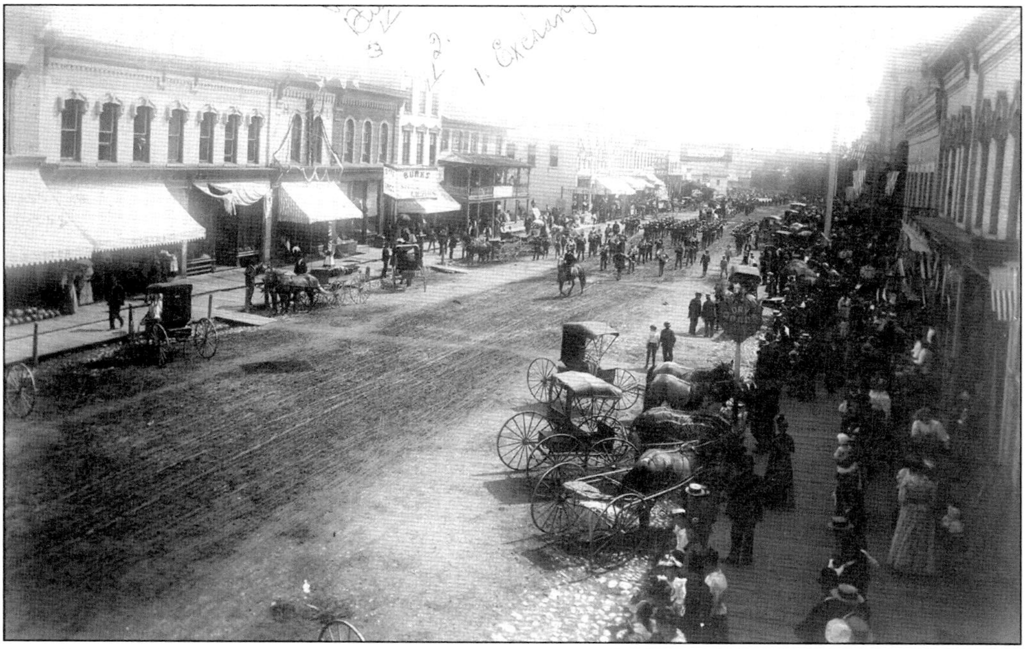

A. V. Burk and Son, top center, was owned and operated by Albert Burk and his son A. Fletcher Burk. The Burks were druggists by trade, but they also offered schoolbooks, lamps, physicians' supplies, and wallpaper. They also had a jewelry department and repaired watches and clocks. The elder Burk came to Wisconsin in 1859 and established this shop in 1871. (WHS.)

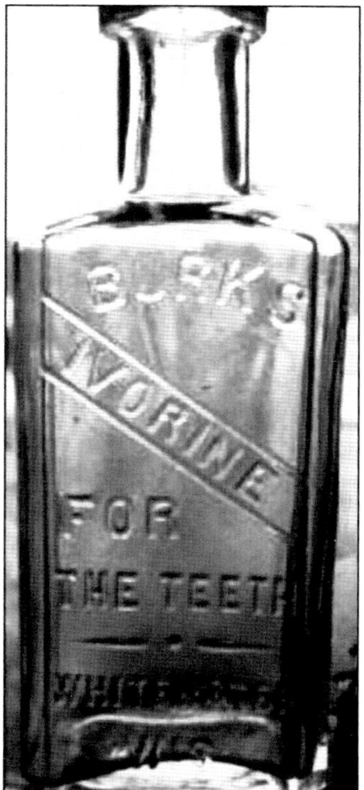

This four-and-one-quarter-inch glass bottle was used to hold Burk's Ivorine for the Teeth, an early toothpaste created by Albert Burk. Toothpaste tubes did not come into use until 1892. The base of the bottle is marked with "WT&CO," and was manufactured by the Whitall Tatum Company in New Jersey. (MrBottles.com.)

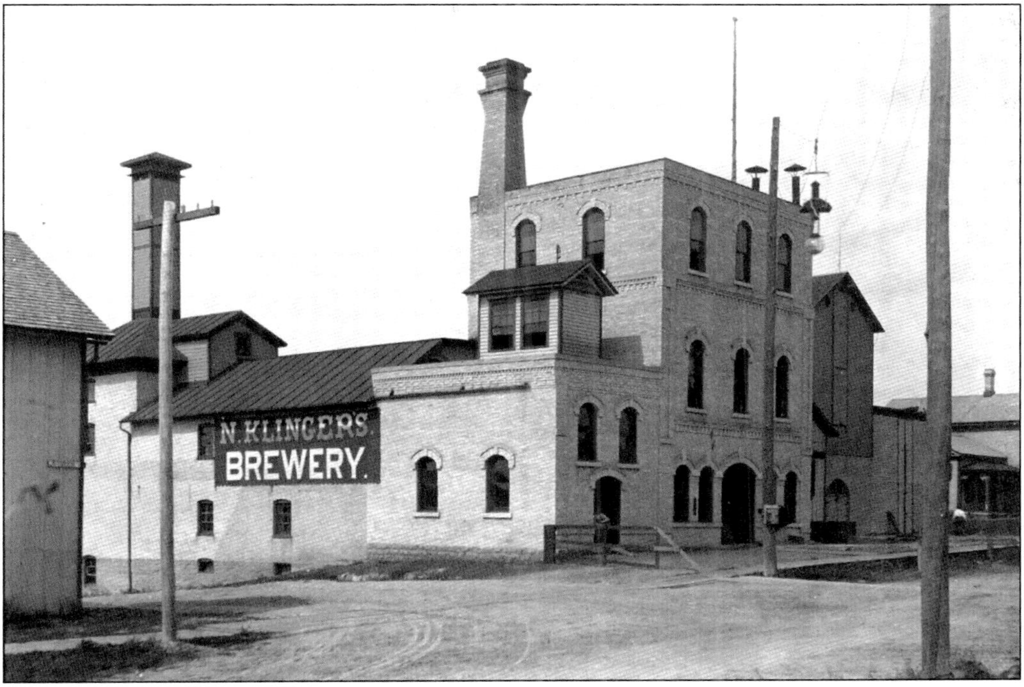

This photograph shows the Nicholas Klinger brewery, established in 1864 on the corner of North and Jefferson Streets. As of 1880, his brewing capacity was 2,000 barrels per year. Small amounts of ale were also brewed here. Klinger was held for trial in federal court in 1879 for his failure to stamp his beer kegs. (Irvin L. Young Memorial Library; Fran Achen Collection.)

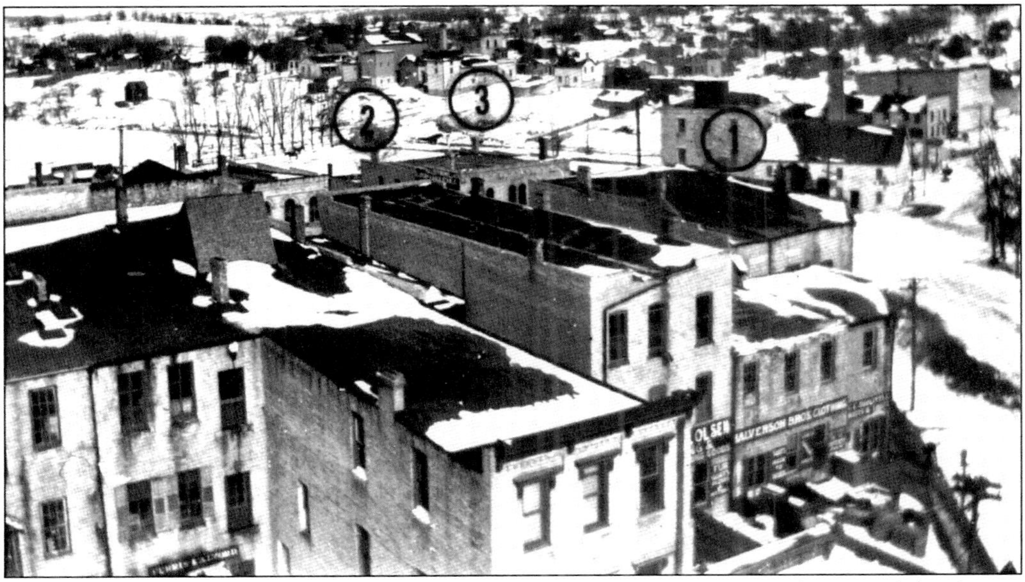

In 1907, the Whitewater Brewery Company No. 3 began operations at the site of the former Klinger Brewery. It brewed and bottled Cream Top, known as the "Beer that Made Milwaukee Furious." Whitewater became a dry town in 1917, and the brewery was closed until 1933. It then reopened, closing for good in 1942. (Author's collection.)

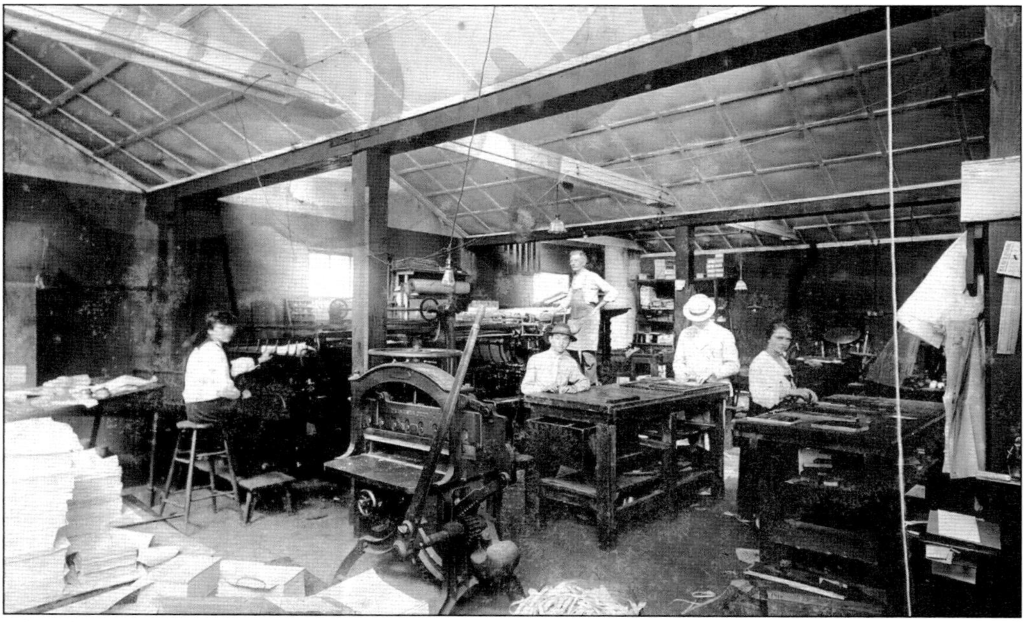

This photograph shows the office of the *Whitewater Register* at the corner of Main and Whitewater Streets. The first issue of the *Whitewater Register* was published in March 1857, and Hannibal L. Rann was the paper's first notable editor. Edwin D. Coe purchased the newspaper in 1871. Coe became the chair of the Wisconsin Press Association and remained publisher of the *Whitewater Register* until his death in 1909. (WHS.)

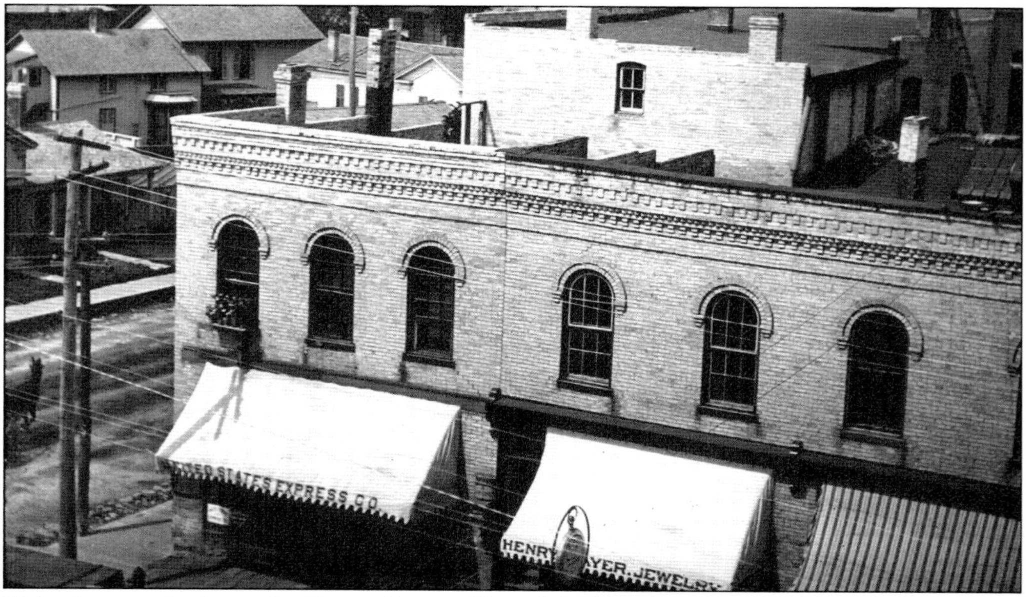

This photograph shows the Center Street office of the U.S. Express Company in 1902. Express companies were an early package delivery service, with agents in every city. Sheldon C. Hall operated a dry goods shop before becoming the city's first Express Company agent by 1861. He handled more than 2,000 incoming and 2,500 outgoing packages a year. (Irvin L. Young Memorial Library: Fran Achen Collection.)

Two

The Esterly Harvesting Machine Company

Of all the factories that made their home in Whitewater, none was as influential as the Esterly Harvesting Machine Company. Its reapers were among the best in the world, consistently winning premiums for quality and performance. The reaper factory owned by George W. Esterly was the most significant Whitewater ever knew. When Esterly departed in late 1892, bringing many employees with him to his new manufacturing facilities in Minneapolis, Whitewater's population went into a downward spiral from which it took decades to recover.

George Esterly was born in 1809 in Ulster County, New York, to Peter Esterly and Rachel Griffith. Esterly worked on early business ventures in brooms, lumbering, and shoe blacking, moving to Alleghany County and then Detroit. In July 1837, Esterly traveled to Milwaukee. After finding the cost of land to be too high in Milwaukee, he ventured into the country, eventually purchasing a claim in Heart Prairie, seven miles southeast of Whitewater. Esterly's farm remained profitable until 1842 and 1843, when the lack of profit from wheat harvests caused him to think of "some easier and cheaper way to harvest wheat." Esterly traveled to St. Louis and purchased castings for several different mechanical harvesters, which he then built in his barn. His first functioning machine was finished in June 1843, and Esterly deemed it a "total failure." As a result, he decided to develop his own "new and independent machine," an improvement on the machines that were currently on the market.

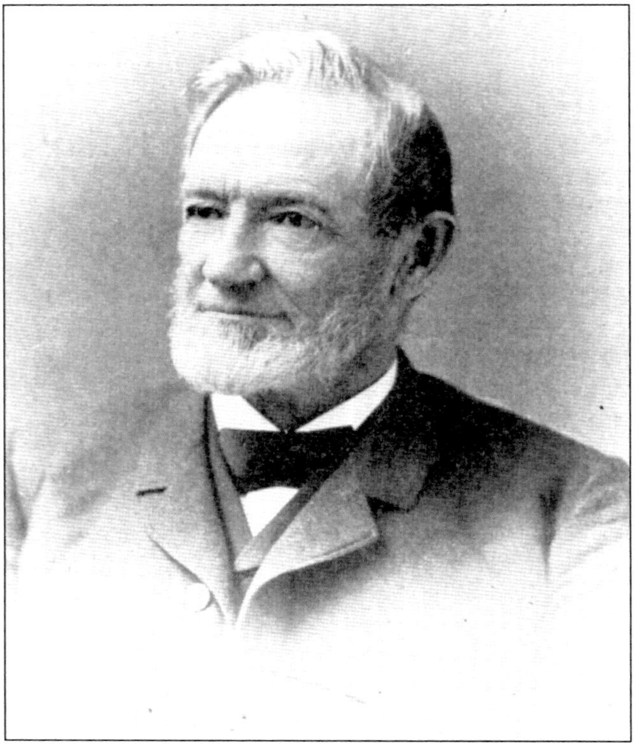

George Esterly, pictured, was not just a reaper manufacturer. He was deeply interested in the currency problems of his time and self-published several financial pamphlets, including *A Plan for Funding the Public Debt and a Safe Return to Specie Payment*, published in 1875, and *A Review of the National Bank System, as to How and Why it Should be Continued*, published in 1887. (WHS.)

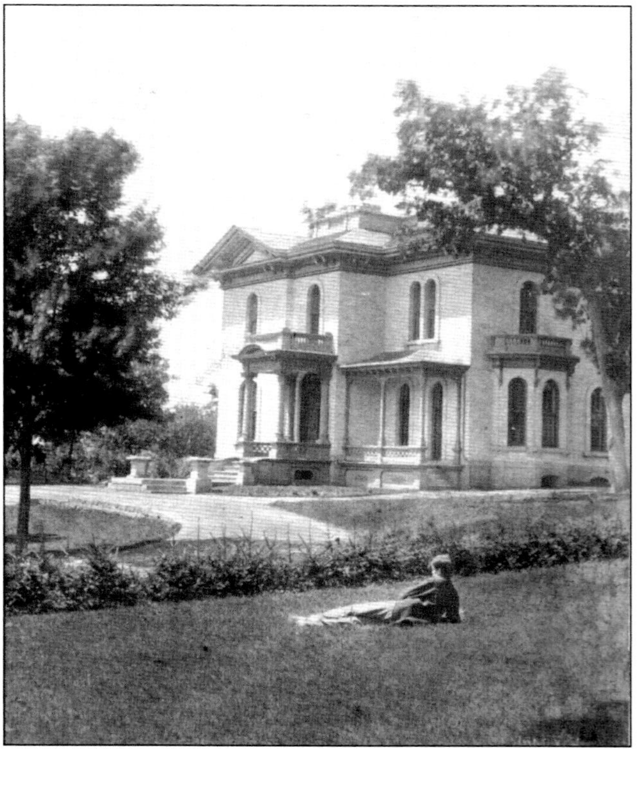

The Esterly home at the end of Center Street was originally built for Judge Eleazer Wakeley. Wakeley ran for state attorney general in 1863, making many enemies in Whitewater. The *Whitewater Register* accused Wakeley of being unpatriotic and characterized his campaign speeches as a "spectacle of . . . political and mental degradation." Wakeley soon left town. Esterly purchased the home in 1864 and remained there for 12 years. (WHS.)

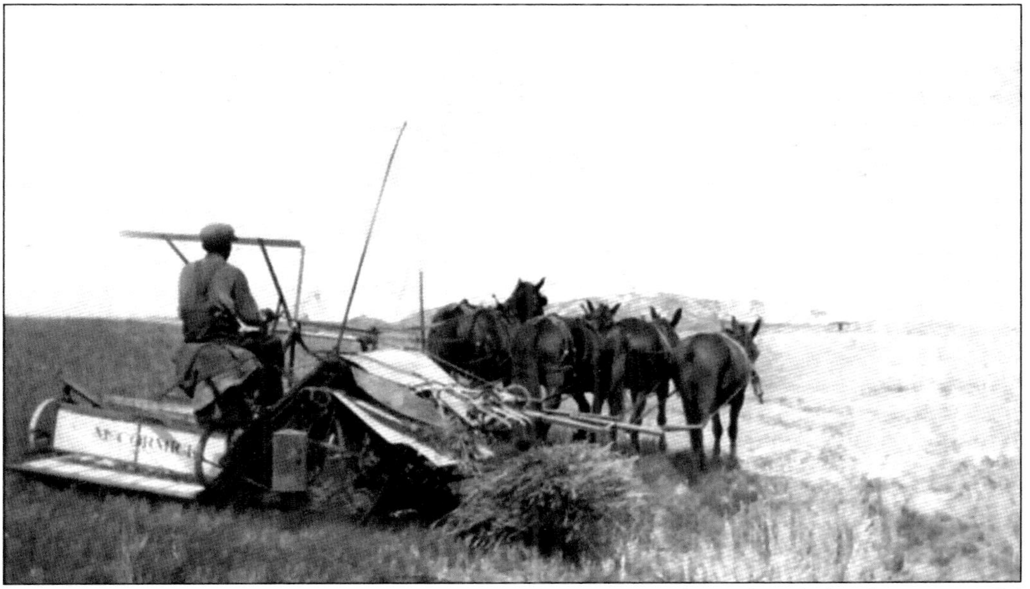

Nearly all scholars acknowledge that Cyrus Hall McCormick produced the first successful mechanical reaping machine. McCormick's machine was patented on June 21, 1834. The first McCormick machine cut stalks of wheat as a team of workers followed, binding the freshly cut stalks of wheat. McCormick's first reaper only required two men to operate, and it could cut as much wheat in a day as 12 to 16 men with handheld reaping hooks. (Jeanne Ringland.)

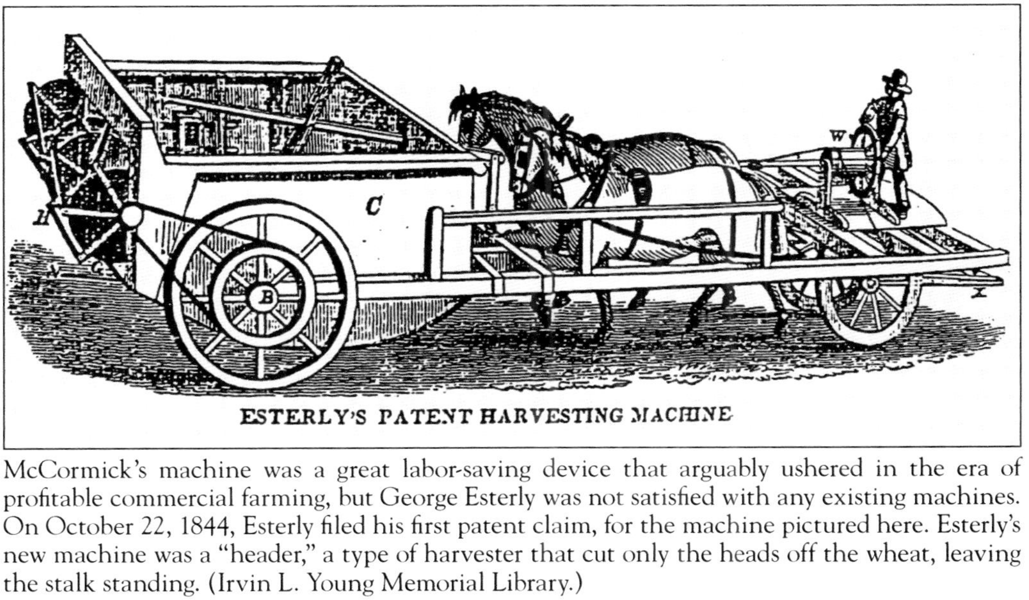

McCormick's machine was a great labor-saving device that arguably ushered in the era of profitable commercial farming, but George Esterly was not satisfied with any existing machines. On October 22, 1844, Esterly filed his first patent claim, for the machine pictured here. Esterly's new machine was a "header," a type of harvester that cut only the heads off the wheat, leaving the stalk standing. (Irvin L. Young Memorial Library.)

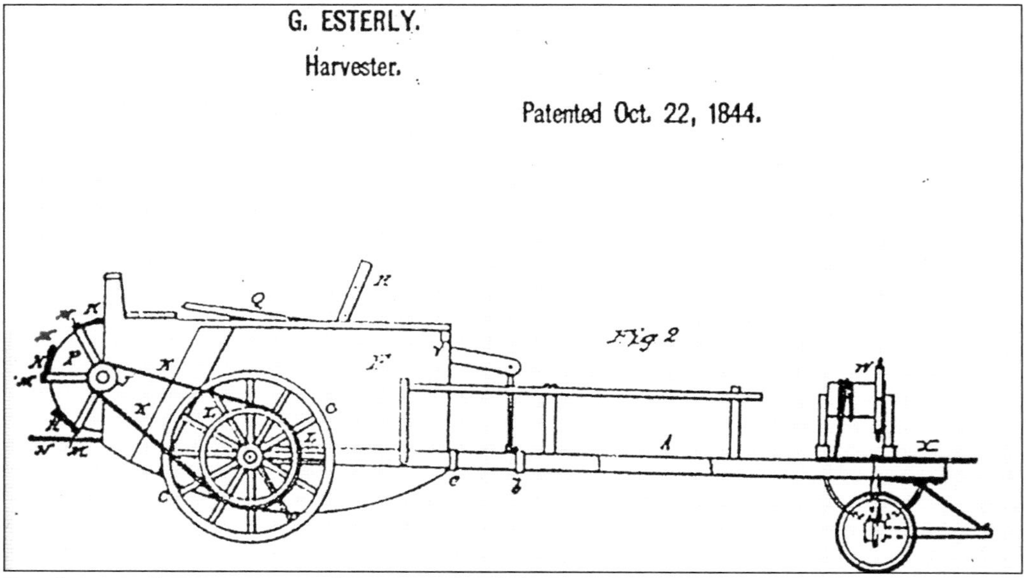

Esterly soon began to manufacture and sell his new harvester, illustrated here. He also manufactured farm wagons and plows, for the time being, and developed a new corn cultivator. While Esterly's first reaper worked to some extent, it was far from perfect. A farmer's letter in an 1846 issue of the *Prairie Farmer* said that Esterly's machine "can cut a field all to destructive smash—it walks over the ground like an elephant." (Irvin L. Young Memorial Library.)

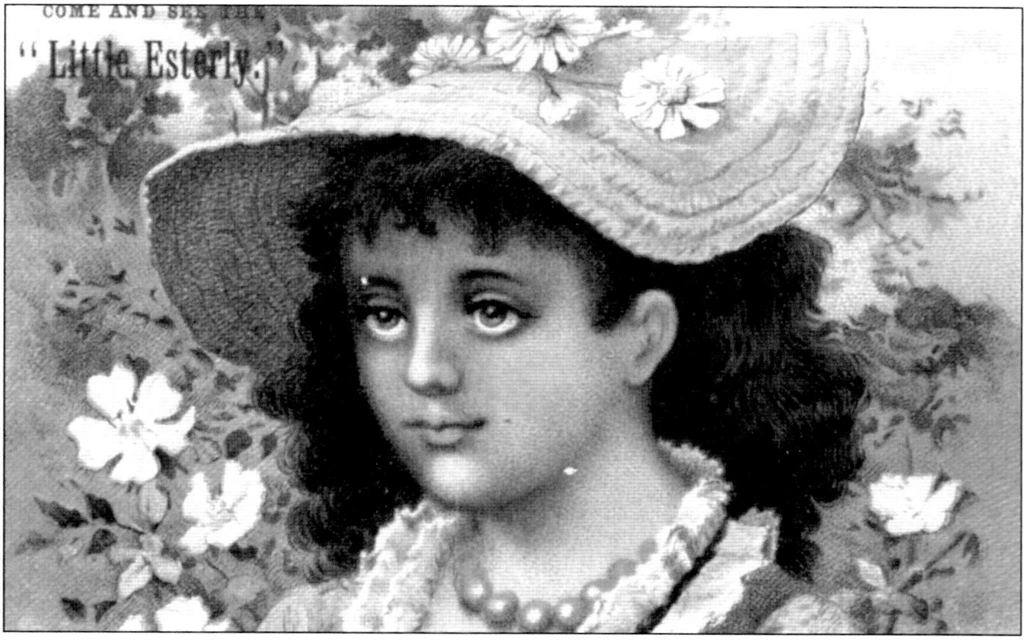

Pictured in this early Esterly advertisement is "Sun Bonnet Sue." Esterly's product was marketed as a competitor to McCormick, and the rivalry came to a head in 1848, when Esterly entered his reaper at the Second Annual Fair of the Chicago Mechanics Institute. Esterly won the gold medal in the harvester competition, besting McCormick's model. However, the decision was extremely controversial. (Irvin L. Young Memorial Library.)

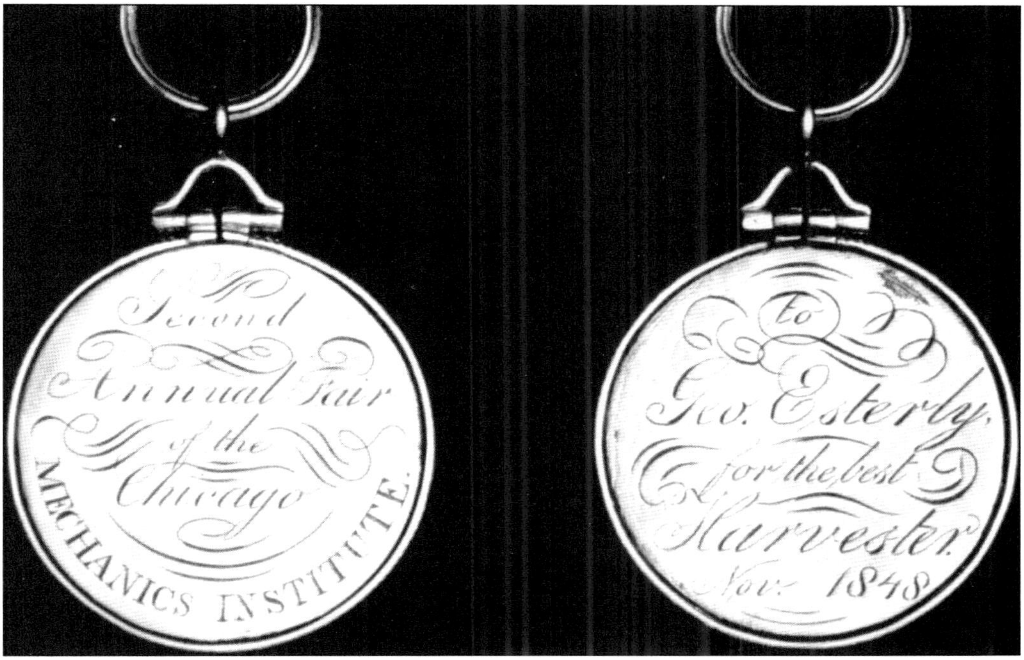

The awarding committee in Chicago claimed no particular knowledge of reapers and characterized itself as "utterly incompetent to decide upon the relative merits of those machines, from seeing them move in a street." The final justification for giving Esterly the award, pictured here, was the fact that "McCormick's machine is more in demand than all others united, and he can take care of himself—therefore, Harvesting Machines, Mr. Esterly, Gold Medal." (Irvin L. Young Memorial Library.)

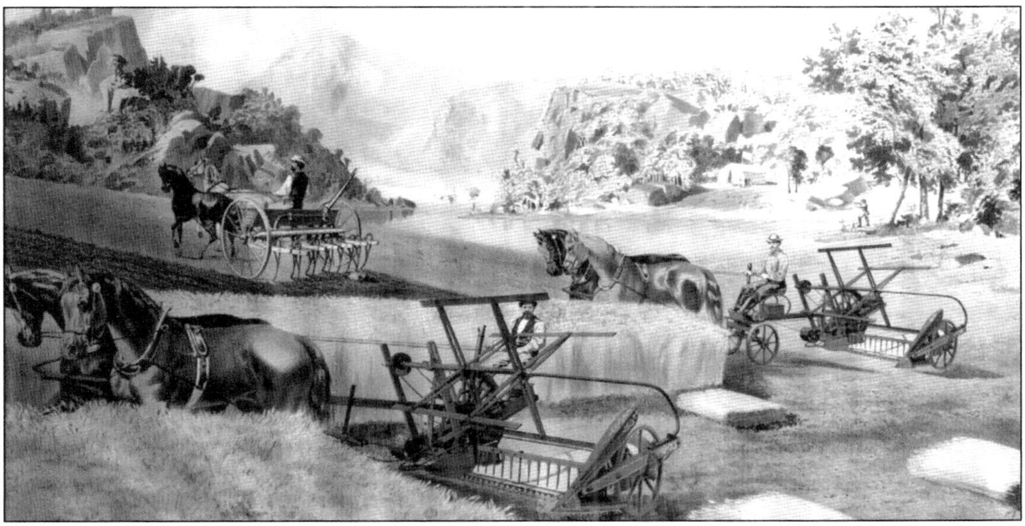

Cyrus Hall McCormick had ample cause to question the committee's decision. In a bitter and sarcastic letter widely published in the Midwest, McCormick dismissed the supposed improvements made by Esterly as "inferior in practice to the construction now used by [McCormick]." Esterly replied, "I wonder whether I could sell Mr. McCormick a few Virginia Reapers, as I am frequently offered them at half price in exchange for my harvester." (WHS.)

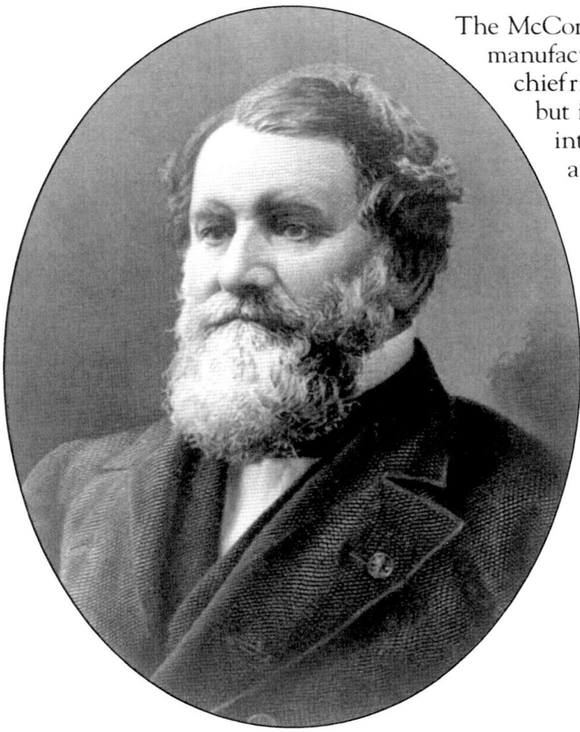

The McCormick-Esterly feud was intense, but both manufacturers managed to coexist. Esterly was a chief rival of Cyrus Hall McCormick (pictured), but in 1890, Esterly's company was absorbed into the short-lived American Harvester, a massive combination of manufacturers run by McCormick's son. Although it would dissolve less than a year later, the McCormicks and Esterlys were, for a short time, allies. (Wikipedia.)

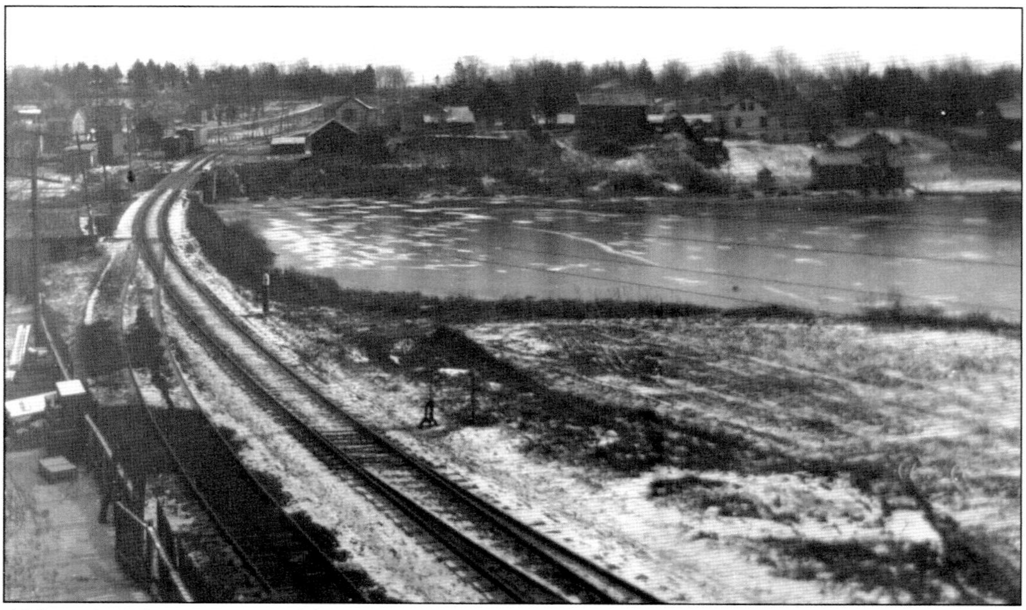

Although Whitewater would come to be defined by Esterly's manufacturing facilities, Esterly originally had a deal in place to purchase land in Hebron. This deal fell through, as did a plan to build a factory in Eagle. Whitewater finally gained Esterly's attention, in part due to a plan to build a railroad from Chicago to Lake Superior that would run just east of Whitewater. (Irvin L. Young Memorial Library; Fran Achen Collection.)

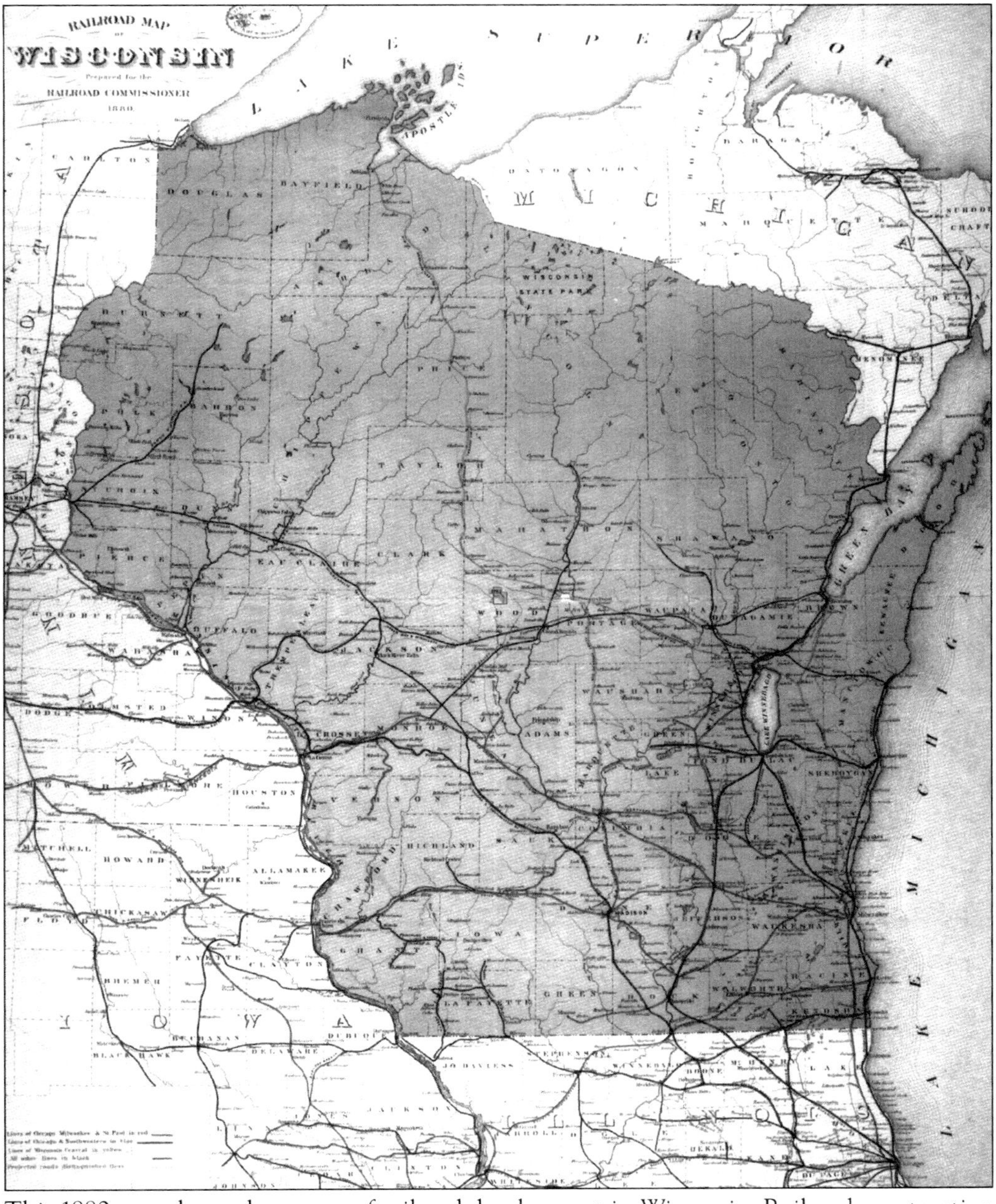

This 1880 map shows the extent of railroad development in Wisconsin. Railroad construction in Wisconsin was slow at first, and railroad development was nearly nonexistent after the Civil War, with not a single mile of track laid from 1865 to 1867. However, construction increased after the development of the Wisconsin Railroad Commission in 1874. But despite the success of Wisconsin railroads, Whitewater was still unable to obtain a second line. (Wisconsin Historical Society, image No. 40157.)

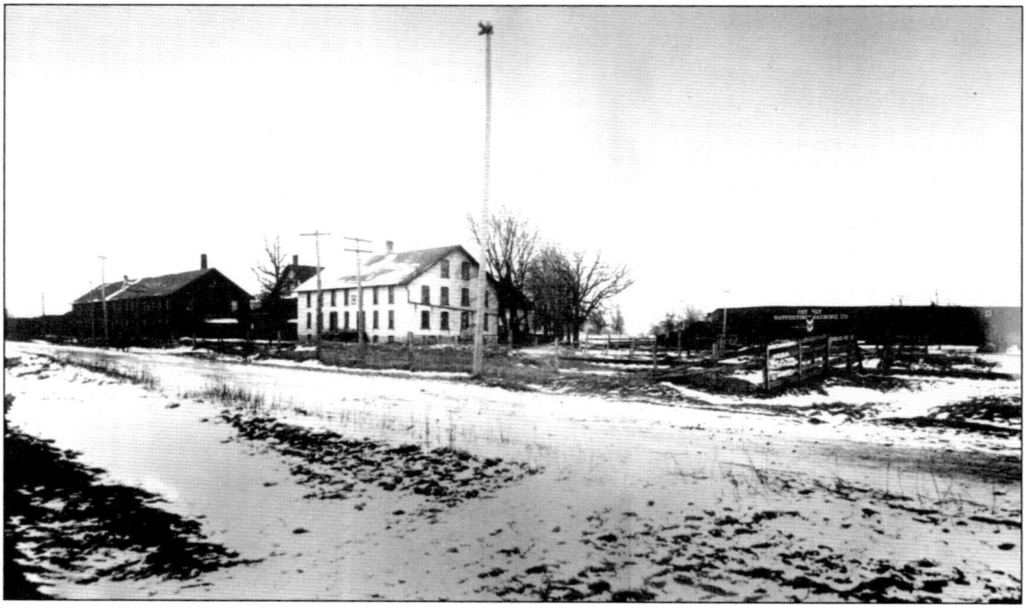

It was not long before Esterly's factory became the central feature of the town. The east side of town became known as Reaperville. An 1857 issue of the *Whitewater Register* states that Esterly's machine "has to be seen to be appreciated" and that it was "acknowledged to be the most complete article of the kind now in use." The factory itself was the most significant in Walworth County. (Irvin L. Young Memorial Library; Fran Achen Collection.)

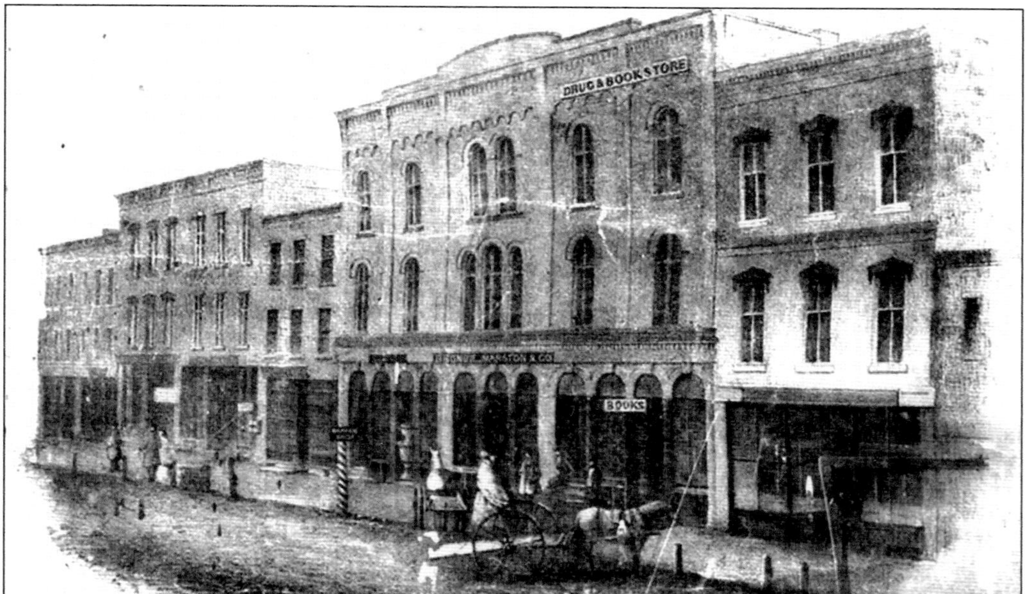

In the early years of the Esterly Company, George Esterly's office was located upstairs above B. A. Jenkins dry goods store on the south side of the first block of Main Street. This building that Esterly occupied burned during the late 1860s, and so subsequent pictures of this building depict the rebuilt sections of the Emporium Block. (University of Wisconsin–Whitewater, Andersen Library.)

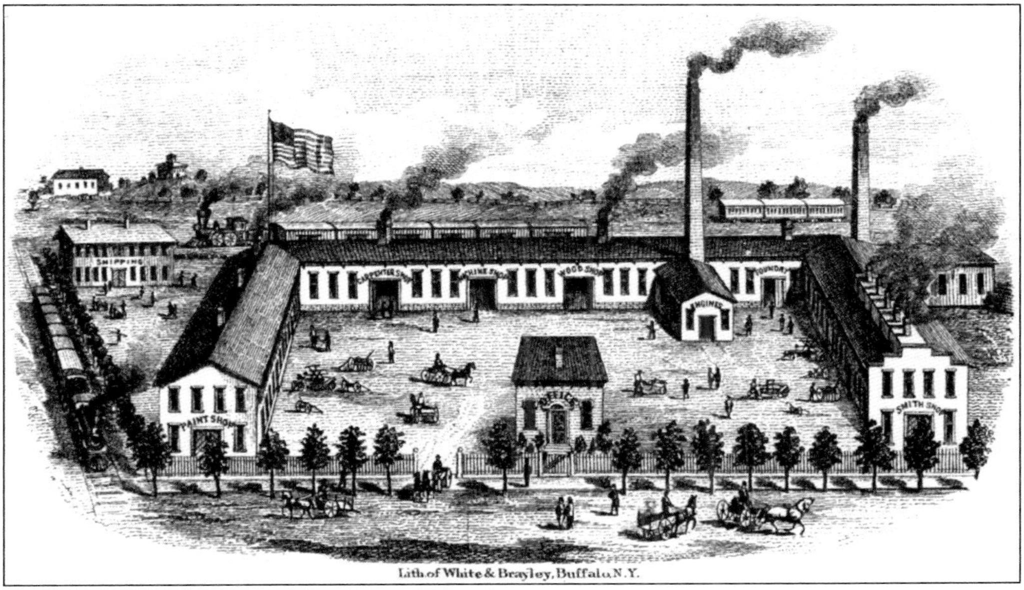

This lithograph shows the Esterly factory. A 40-by-100-foot foundry building contained a foundry as well as a blacksmithing department, in which Esterly manufactured the sickles for his machine with steel imported from England. There was also a 40-by-100-foot machine shop, with machines for sawing and shaping wood, as well as machines for iron finishing. The reapers were assembled in an 80-by-100-foot, two-story wood shop. (Irvin L. Young Memorial Library.)

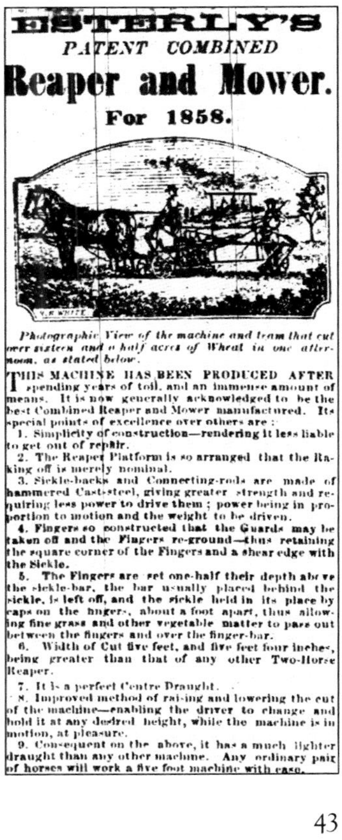

ESTERLY'S

PATENT COMBINED

Reaper and Mower.

For 1858.

Photographic View of the machine and team that cut over sixteen and a half acres of Wheat in one afternoon, as stated below.

THIS MACHINE HAS BEEN PRODUCED AFTER spending years of toil, and an immense amount of means. It is now generally acknowledged to be the best Combined Reaper and Mower manufactured. Its special points of excellence over others are:

1. Simplicity of construction—rendering it less liable to get out of repair.

2. The Reaper Platform is so arranged that the Raking off is merely nominal.

3. Sickle-backs and Connecting-rods are made of hammered Cast-steel, giving greater strength and requiring less power to drive them; power being in proportion to motion and the weight to be driven.

4. Fingers so constructed that the Guards may be taken off and the Fingers re-ground—thus retaining the square corner of the Fingers and a shear edge with the Sickle.

5. The Fingers are set one-half their depth above the sickle-bar, the bar usually placed behind the sickle, is left off, and the sickle held in its place by caps on the fingers, about a foot apart, thus allowing fine grass and other vegetable matter to pass out between the fingers and over the finger-bar.

6. Width of Cut five feet, and five feet four inches, being greater than that of any other Two-Horse Reaper.

7. It is a perfect Centre Draught.

8. Improved method of raising and lowering the cut of the machine—enabling the driver to change and hold it at any desired height, while the machine is in motion, at pleasure.

9. Consequent on the above, it has a much lighter draught than any other machine. Any ordinary pair of horses will work a five foot machine with ease.

Esterly patented a self-raking reaper in 1859, but it was a failure. According to *Wisconsin Farmer*, Esterly spent "many years experimenting, and an expenditure of perhaps $50,000" in coming to the conclusion that it was too difficult to build and maintain successful self-rakers. Esterly successfully repaid all creditors, but only after 10 years of work. (Whitewater Register.)

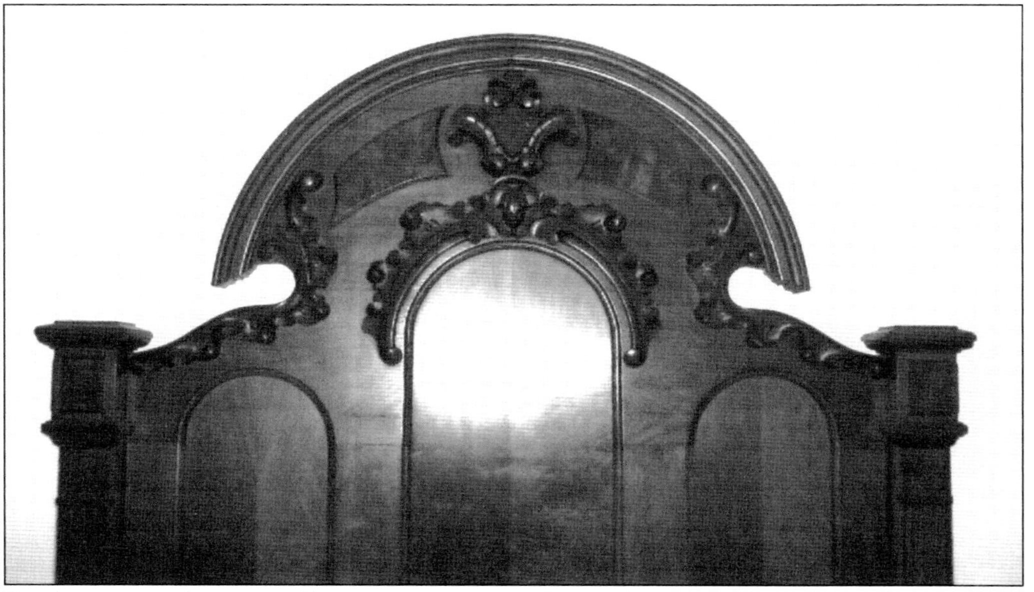

Esterly continued to expand his operations into other outlets. He built seeders, cultivators, and horse rakes. He also found "a ready sale" for "superior furniture" that he manufactured, such as this headboard. Esterly put his saws to work for the general public, offering custom sawing services. He also manufactured sleighs and coffins. The reaper dominated Esterly's business, but he was a creative entrepreneur. (Hulburt family.)

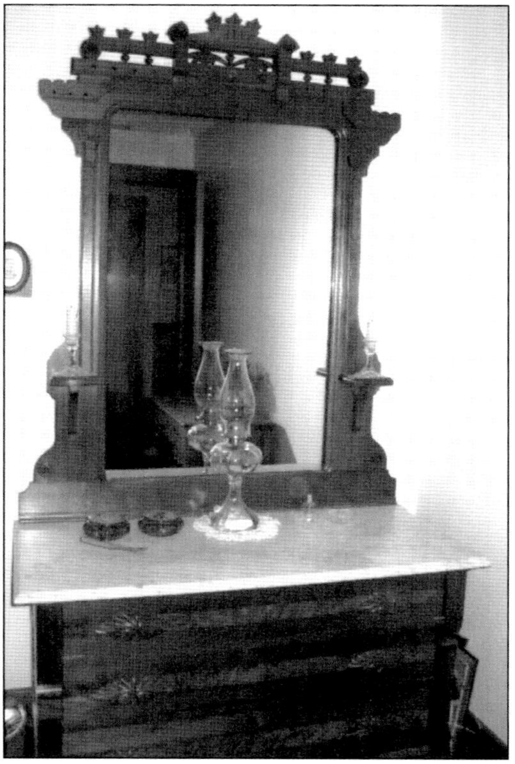

George Esterly began manufacturing furniture in 1864, stopping when his furniture factory burned down in 1867. In the spring of 1874, Esterly organized the Whitewater Furniture Factory in a large, newly built, three-story brick structure. The factory employed as many as 90 people and produced the dresser shown here. Esterly discontinued his furniture manufacturing in 1881. (Hulburt family.)

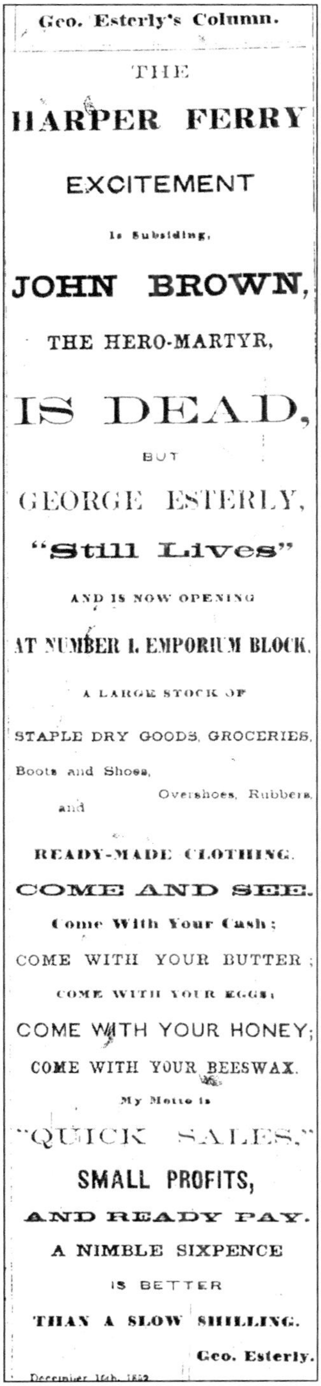

Geo. Esterly's Column.

THE

HARPER FERRY

EXCITEMENT

Is Subsiding,

JOHN BROWN,

THE HERO-MARTYR,

IS DEAD,

BUT

GEORGE ESTERLY,

"Still Lives"

AND IS NOW OPENING

AT NUMBER 1, EMPORIUM BLOCK,

A LARGE STOCK OF

STAPLE DRY GOODS, GROCERIES,

Boots and Shoes,

Overshoes, Rubbers,

and

READY-MADE CLOTHING.

COME AND SEE.

Come With Your Cash;

COME WITH YOUR BUTTER;

COME WITH YOUR EGGS;

COME WITH YOUR HONEY;

COME WITH YOUR BEESWAX.

My Motto is

"QUICK SALES,"

SMALL PROFITS,

AND READY PAY.

A NIMBLE SIXPENCE

IS BETTER

THAN A SLOW SHILLING.

Geo. Esterly.

Always looking for new ways to expand his business, George Esterly also operated a short-lived dry goods store on the first floor of the Emporium building. This 1859 advertisement announces the opening of this store, offering "quick sales, small profits, and ready pay." The store sold groceries and other dry goods before closing in 1862. (Whitewater Register.)

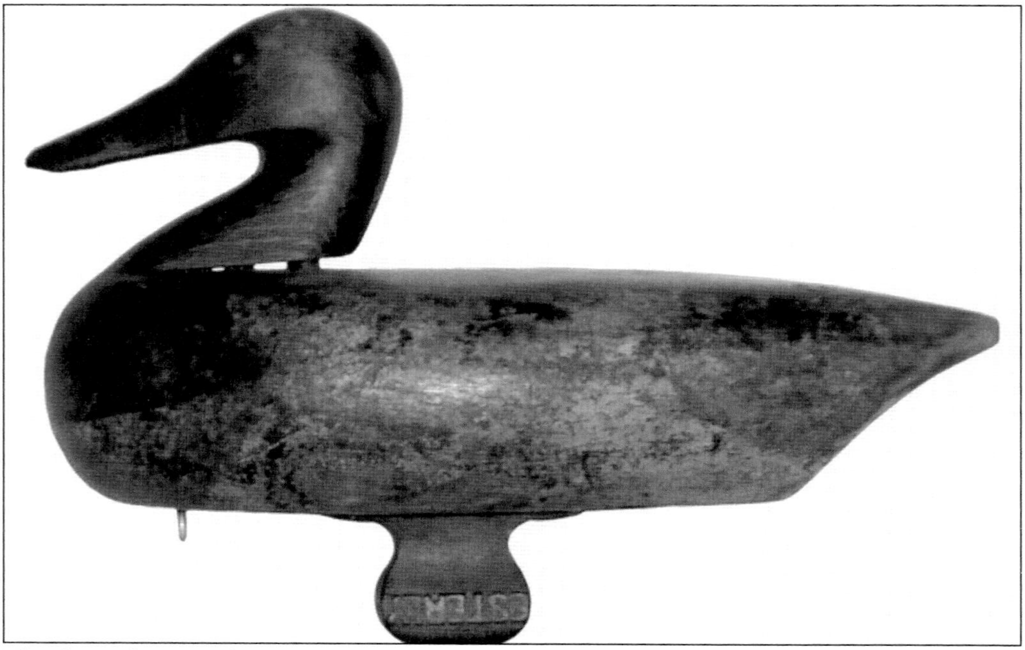

The diversification of Esterly's products may even have extended as far as duck decoys. George Esterly's son, George W., was a member of the Blackhawk Club, a shooting club located on the eastern end of Lake Koshkonong. The club was incorporated in 1875 with Lucius Winchester as its first president. Lake Koshkonong was a prime area for hunting canvasback ducks during the late 1800s. (Bob Lehmkuhl.)

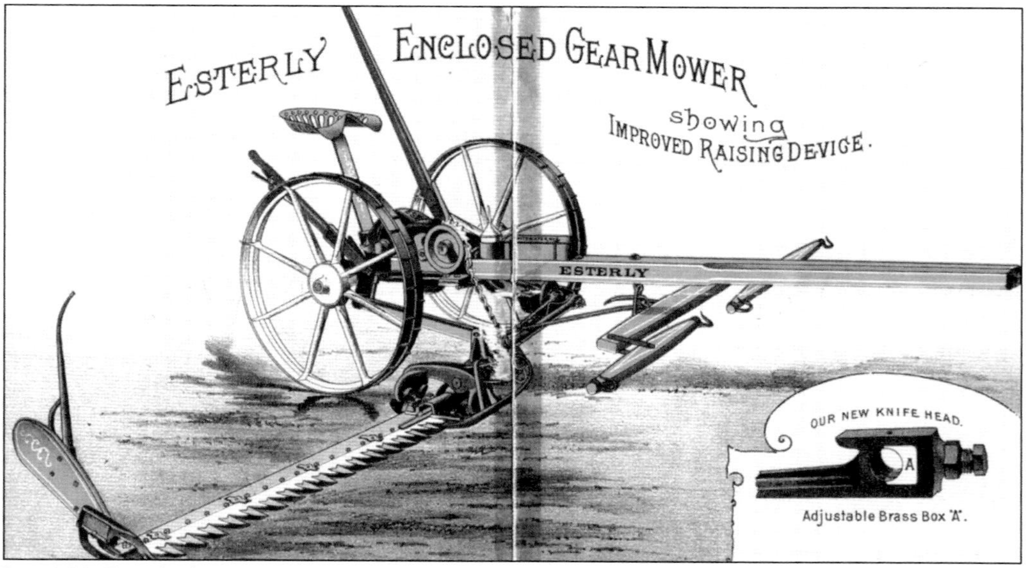

In 1866, Esterly began to manufacture the Broadcast Seeder and Cultivator, a labor-saving device used to spread grain and seeds. More seeders than reapers were built during the 1870s, but the reaper was still in great demand across the country. As far away as Texas, the Esterly machine was "preferred to any machine on the market." Esterly also manufactured the enclosed gear mower, pictured here. (Irvin L. Young Memorial Library.)

At 11:00 on the night of January 4, 1867, a fire consumed Esterly's machine shop, furniture manufactory, and turning establishment (used to shape wood). Rumors immediately began to circulate that Esterly would not rebuild. In June 1867, the *Janesville Gazette* wrote that Esterly sought a more central location than Whitewater and was considering Janesville as the location for a new factory. Citizens feared that "if Mr. Esterly does not rebuild here, houses and lots will be plenty as blackberries, and nobody to buy." The *Whitewater Register* wrote that Whitewater without Esterly would turn into "a quiet rural retreat, where the grass shall grow up in our streets unobstructed by teams, and the cow and the swine will pasture pleasantly in the village." Fortunately, Esterly resumed production in Whitewater by the end of the year. (Irvin L. Young Memorial Library; Fran Achen Collection.)

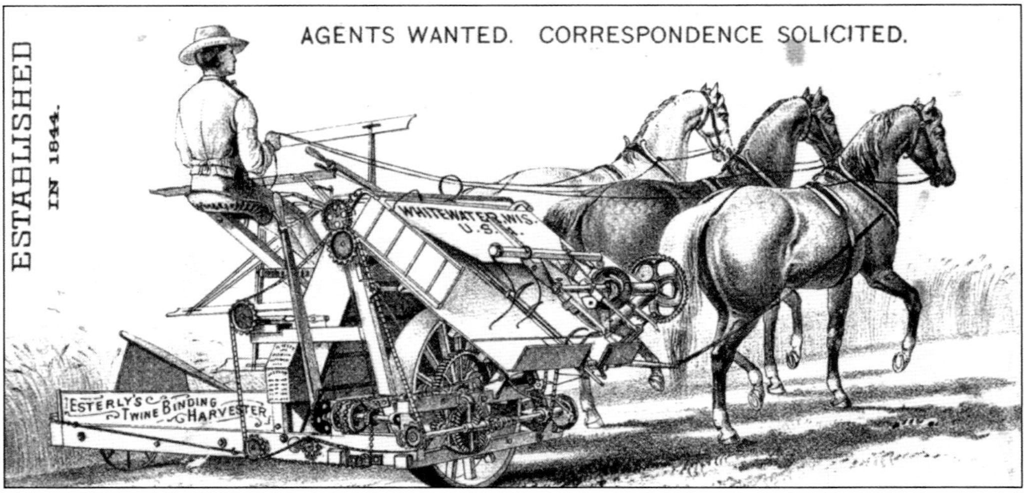

The Beloit Grain-Binder, invented by John F. Appleby, provided a necessary breakthrough in the binding of harvested wheat. This twine-binding machine was exhibited at the 1877 Wisconsin State Fair and described as "perfect" by a *Chicago Daily Tribune* correspondent. Appleby came to Whitewater in 1878 to work with George Esterly on improving his binding device, pictured here as produced by Esterly. (Irvin L. Young Memorial Library.)

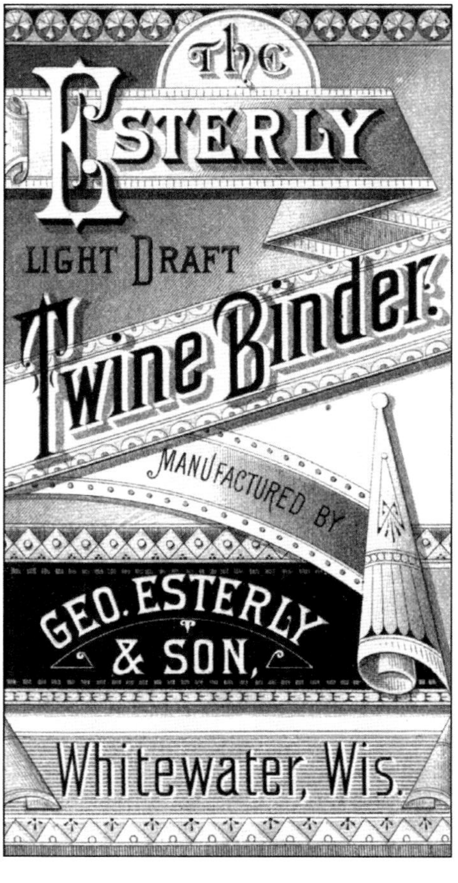

This 1888 promotional pamphlet was handed out not only as an advertising piece that extolled the virtues of Esterly products but as a place to record notes and information in a "diary" of sorts. It contains a calendar and a list of dealers from Maine to Montana that carried Esterly products. (Irvin L. Young Memorial Library.)

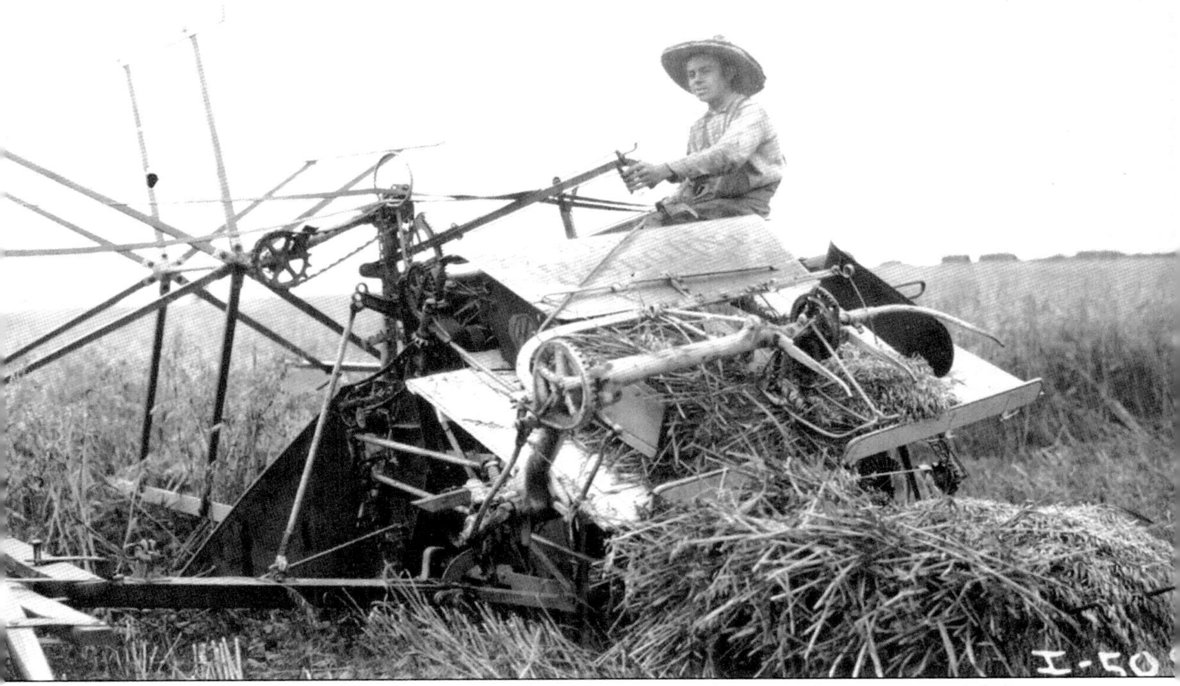

In June 1883, Esterly faced a serious legal challenge. D. M. Osborne and Company, of Auburn, New York, was manufacturing a self-binder (pictured here) that was similar to Esterly's model. Esterly flooded Wisconsin with circulars, claiming patent infringement and threatening to prosecute Osborne and anyone who purchased his product. Osborne filed suit himself, claiming that Esterly's circulars were untrue and issued maliciously, with intent to harm Osborne's business. Osborne sought damages of $150,000. The *New York Times* described the case as "one of the most important ever recorded in the local courts," but it came to a rather anticlimactic conclusion. After a prolonged battle, the suit was discontinued on November 21, 1884. No decision was reached, and neither party was required to pay any costs. George Esterly had dodged a potential disaster. (Wisconsin Historical Society, image No. 44823.)

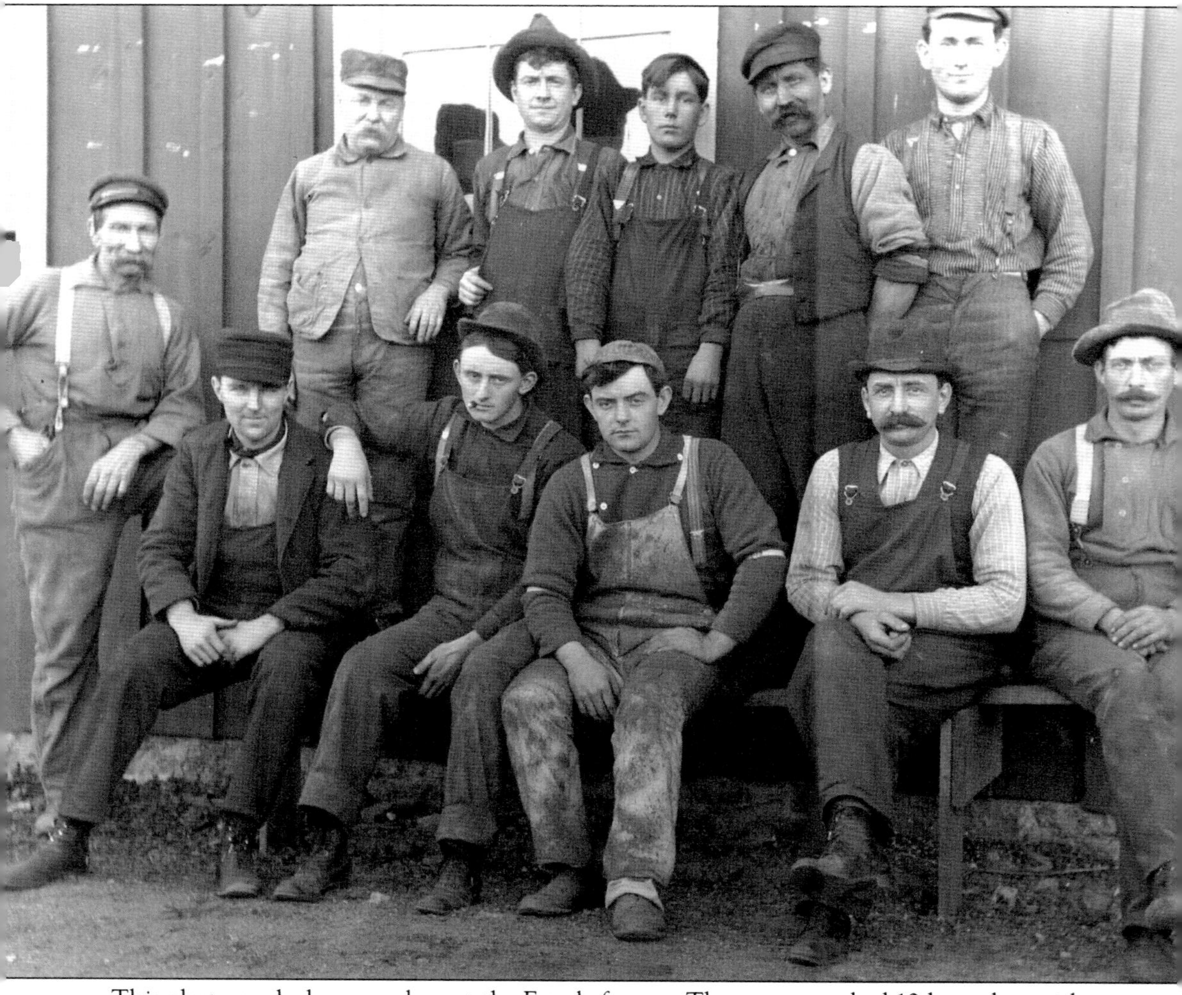

This photograph shows workers at the Esterly factory. These men worked 10-hour days with one half-hour break for dinner. The Esterly Harvesting Machine Company was very concerned with worker safety. In an 1888 report, an inspector Moore expressed concern that employees were "likely to get caught under the elevator, in its descent." Esterly responded by using guards at every elevator well, and foremen were to "discharge any worker who leaves them open." Moore found that a fire escape should be built for the female employees working on the third floor. Esterly built three. Safety guidelines were not as strict as in modern times, but Esterly more than followed all regulations, often going far beyond what the state requested. A state inspection in April 1889 found that "everything ordered or suggested at former inspections has been strictly carried out." (Irvin L. Young Memorial Library; Fran Achen Collection.)

Three

THE ESTERLY COMPANY'S DEPARTURE AND FAILURE

While Whitewater allowed George Esterly to develop a profitable business, Esterly repeatedly sought advances in transportation to benefit not only his company but the town as a whole. Esterly's business was thriving in Whitewater, but additional railroads would have allowed him to sell his reapers to even more markets. The company was a manufacturing power in its own right, and it could have reached even greater heights with better railroad connections. His unsuccessful railroad advocacy frustrated Esterly and ultimately proved to be a huge factor in his decision to move the company to a larger city that offered greater opportunities.

In 1892, the Esterly Company closed its operations in Whitewater and moved to Minneapolis. But why did Esterly decide to uproot his thriving business? For one thing, Minneapolis offered opportunity. Esterly's regular lobbying for a new railroad indicates dissatisfaction with Whitewater's transportation facilities. Minneapolis provided better railroads and shipping outlets. Also, a great deal of capital was available in Minneapolis, and Esterly's business had the potential to significantly expand. Esterly's move was also the product of a shifting frontier. His fate was tied to that of the frontier wheat farmer, the chief consumer of his reaper, who sought cheaply available land. The cheapest land could be had on the wheat frontier, which was gradually shifting to the West.

Minnesota itself was a natural outlet for Esterly. Wheat-growing in Minnesota grew from 18,866,073 bushels in 1870 to 52,300,247 bushels in 1890. At the same time, diversified farming overtook Wisconsin, and its days as a dominant power in wheat were numbered. The business logic of Esterly's move was clear, but his factory would not survive for long in Minneapolis.

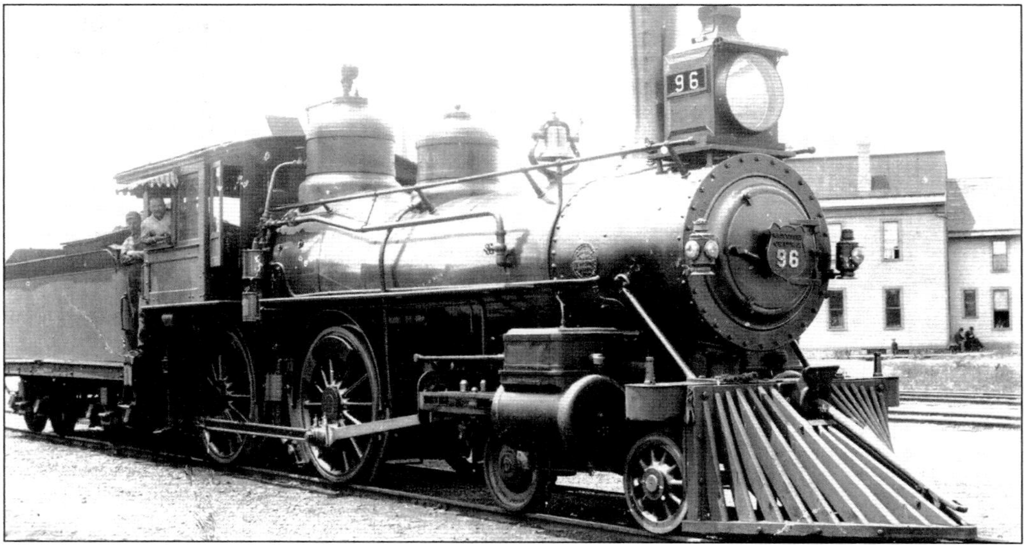

George Esterly was a founding director of the Portage and Superior Railroad that eventually became the Wisconsin Central. He spent much time lobbying the state legislature in Madison for a second railroad for Whitewater, and his unsuccessful railroad advocacy spanned four decades. This photograph shows an engine from the Wisconsin Central, which connected Minnesota and Chicago by way of Waukesha. (Wisconsin Historical Society, image No. 24321.)

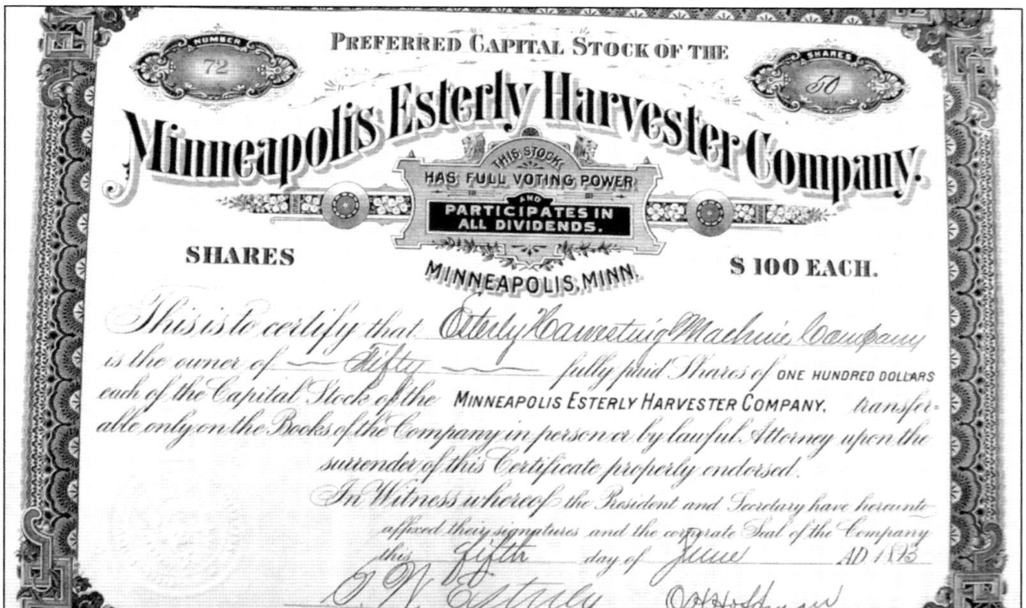

NUMBER 72

PREFERRED CAPITAL STOCK OF THE

SHARES 50

Minneapolis Esterly Harvester Company.

THIS STOCK HAS FULL VOTING POWER AND PARTICIPATES IN ALL DIVIDENDS.

SHARES

MINNEAPOLIS, MINN.

$ 100 EACH.

This is to certify that Esterly Harvesting Machine Company is the owner of Fifty fully paid Shares of ONE HUNDRED DOLLARS each of the Capital Stock of the MINNEAPOLIS ESTERLY HARVESTER COMPANY, transferable only on the Books of the Company in person or by lawful Attorney upon the surrender of this Certificate properly endorsed.

In Witness whereof the President and Secretary have hereunto affixed their signatures and the corporate Seal of the Company this Fifth day of June AD 189

President. Secretary.

In the early 1890s, the Twin Cities competed to be the western site of the Walter A. Wood Harvester Company. St. Paul collected $2 million in capital and secured Wood's plant, so Minneapolis brought Esterly to town. Industrial investor Thomas Barlow Walker became Esterly's patron. He provided a 10-acre plant, with two three-story brick buildings and the means to employ 500–600 men. Pictured is a stock certificate from the Minneapolis Esterly Harvester Company. (Minnesota Historical Society.)

George Esterly died in Hot Springs, South Dakota, in June 1893. An obituary in the *Farm Implement News* characterized Esterly as "[a] man of never-ceasing mental activity . . . there was scarcely a subject which he had not studied, and upon which he had not radical opinions . . . all were forced to admire even where they could not agree." Pictured is his last home in Whitewater, located on Main Street. (WHS.)

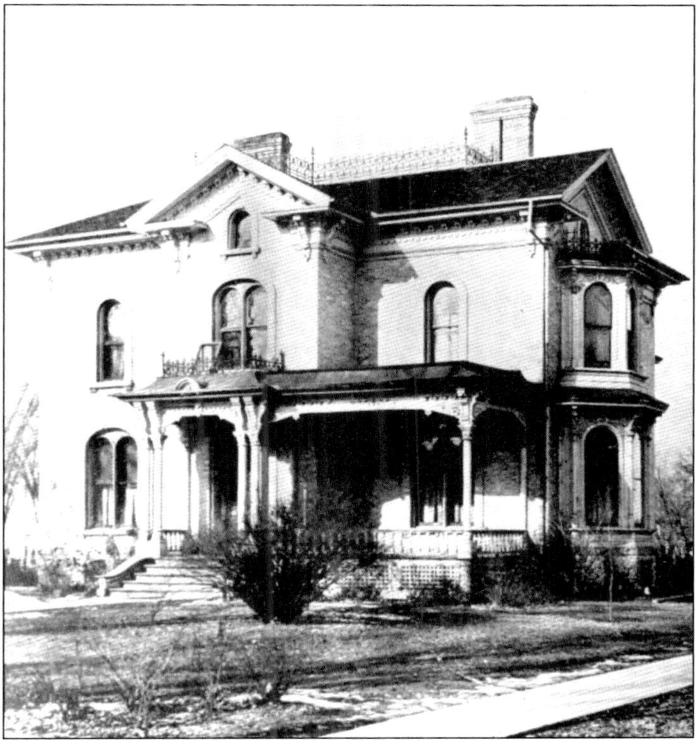

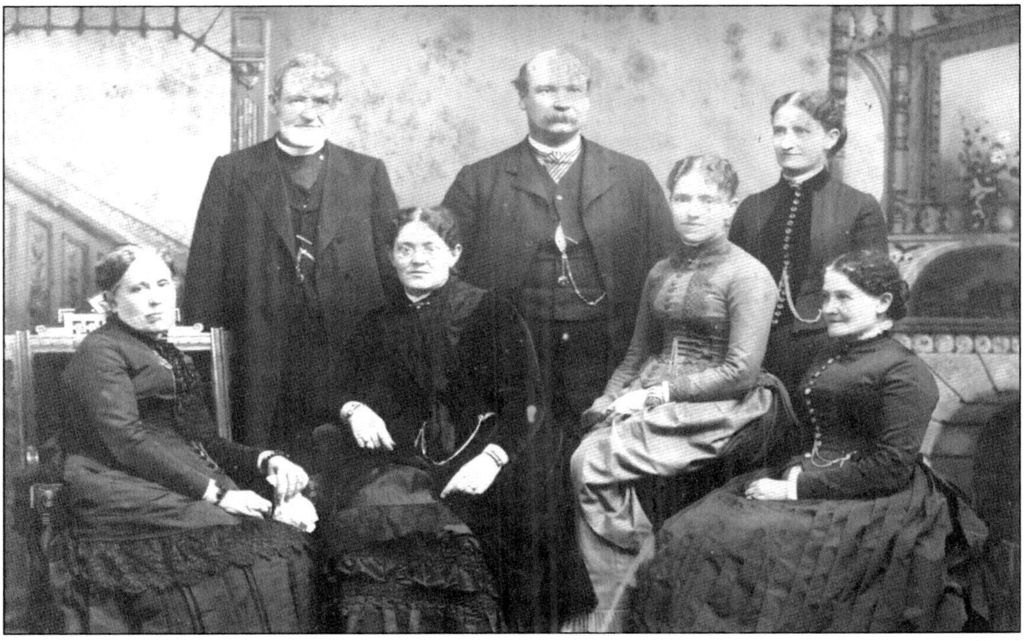

This 1888 photograph shows the Esterly family. Pictured are Caroline Esterly (Esterly's third wife), George W. Esterly, Phoebe Esterly Crites, Eva Esterly Page, Kate Haines Esterly (wife of George W. Esterly), and Caroline Esterly Nicholls. Another Esterly daughter, Mary, is not pictured. George Esterly's first wife, Jane Lewis, was the mother of all his children. (WHS.)

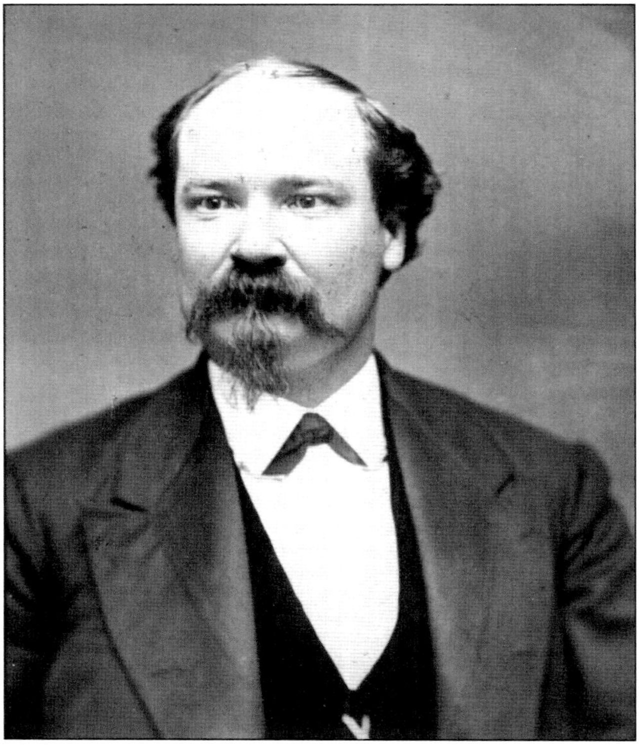

George W. Esterly, pictured, left the University of Wisconsin in 1862 to help straighten out his father's business affairs. He became a full partner in his father's business in 1872. When the company moved to Minneapolis, George W. became president and oversaw the last few years of the company's operations. In 1896, he accepted a position as deputy auditor of the treasury department in Washington, D.C. (WHS.)

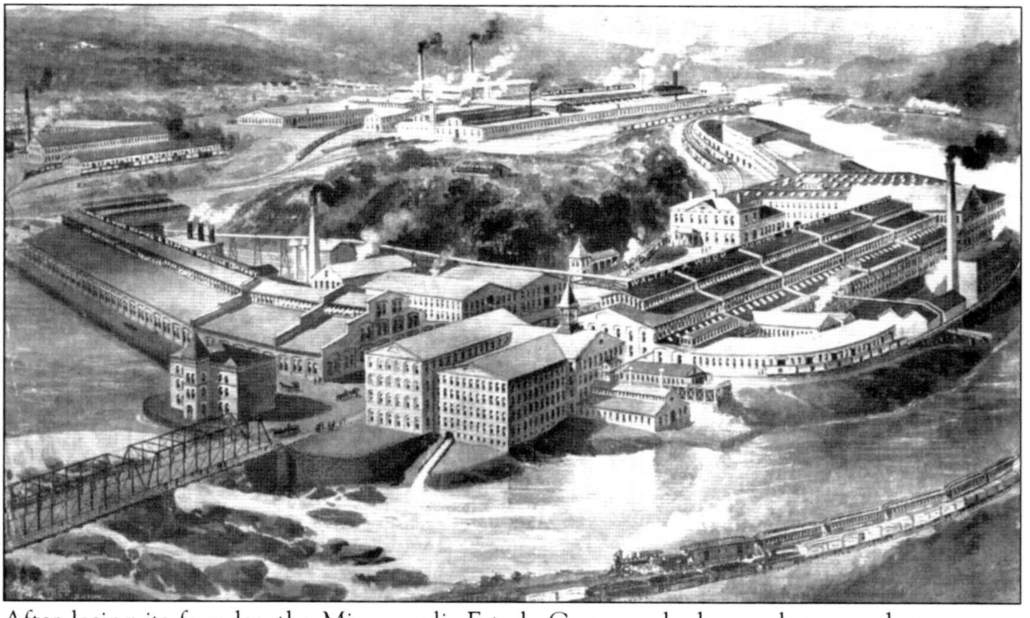

After losing its founder, the Minneapolis Esterly Company had scarcely a month to recover before facing another legal battle. On July 18, 1893, the Walter A. Wood Company, pictured here, brought a $2 million suit against Esterly over the right to manufacture self-binders under the Appleby patents. This was a case of great national importance, as the victor stood to hold an essential monopoly of the reaper trade in the Northwest. (Author's collection.)

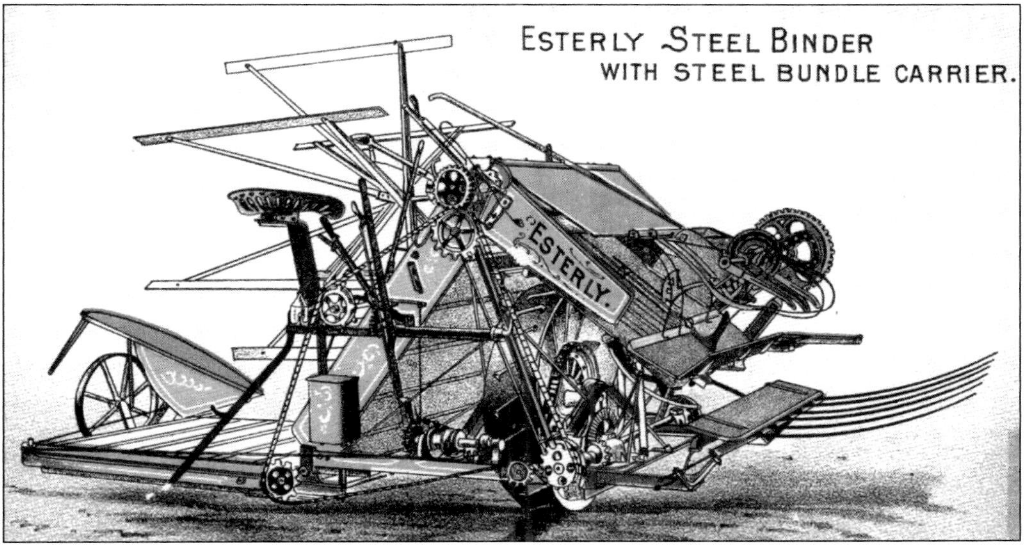

Before this case was resolved, the Esterly Company felt the strain of financial panic. In the August 17, 1893, issue of the *Whitewater Register*, a brief clipping states that the Minneapolis Esterly Company "closed temporarily last Saturday, and all the Whitewater employees except those working by the year, have returned home." At this point, the factory's closure was assumed to be temporary, but this was the beginning of the end. (Irvin L. Young Memorial Library.)

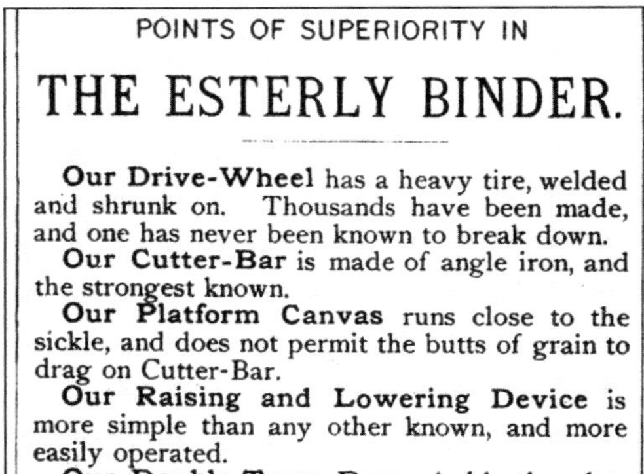

POINTS OF SUPERIORITY IN

THE ESTERLY BINDER.

Our Drive-Wheel has a heavy tire, welded and shrunk on. Thousands have been made, and one has never been known to break down.

Our Cutter-Bar is made of angle iron, and the strongest known.

Our Platform Canvas runs close to the sickle, and does not permit the butts of grain to drag on Cutter-Bar.

Our Raising and Lowering Device is more simple than any other known, and more easily operated.

Our Double Truss Brace holds the platform and elevator frames solid, so that it is impossible for them to get out of square.

Our Canvas Adjusters, at each end of the Binder, help to form a good bundle in long or short grain.

Our Three Discharge Arms prevent dragging of the bundles even in badly tangled grain.

Our Knot-Tying Device has been so simplified, that it astonishes even "old binder men," to see how little machinery is necessary to tie a knot.

The Old Plunger Bolt, Tucker Finger and Slack Arm, with all their springs and com-

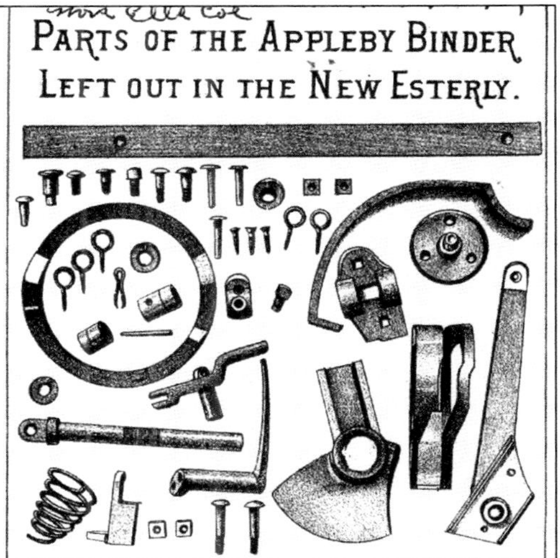

On January 16, 1894, the Wood Company was declared the victor in its suit against Esterly. The Esterly Company was ordered to quit manufacturing its Appleby machine, advertised here, unless a $25,000 indemnity was paid to the Wood Company within 10 days. In the wake of the panic of 1893, the Esterly Company simply did not have this kind of readily available capital. (Irvin L. Young Memorial Library.)

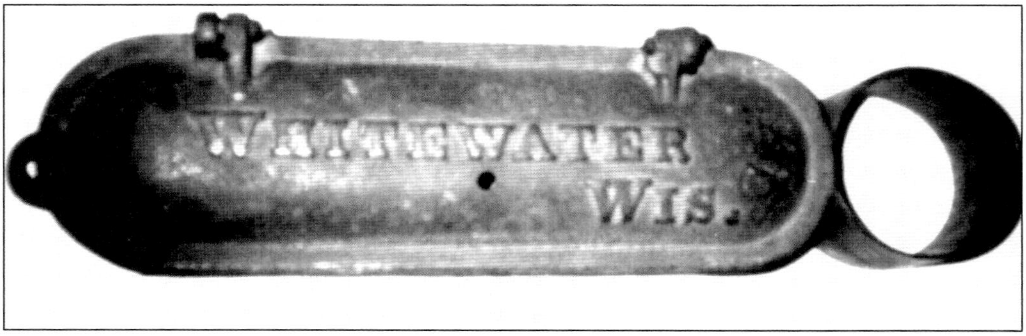

By this point, it was clear that the company would not survive. The company and its creditors set up a debt payment schedule to allow the company to repay its debts slowly over the next three years before the end of its operations. The factory would remain in operation, producing as many reapers as possible to cover outstanding debts. Pictured is a toolbox and oil can holder for the reaper. (Irvin L. Young Memorial Library.)

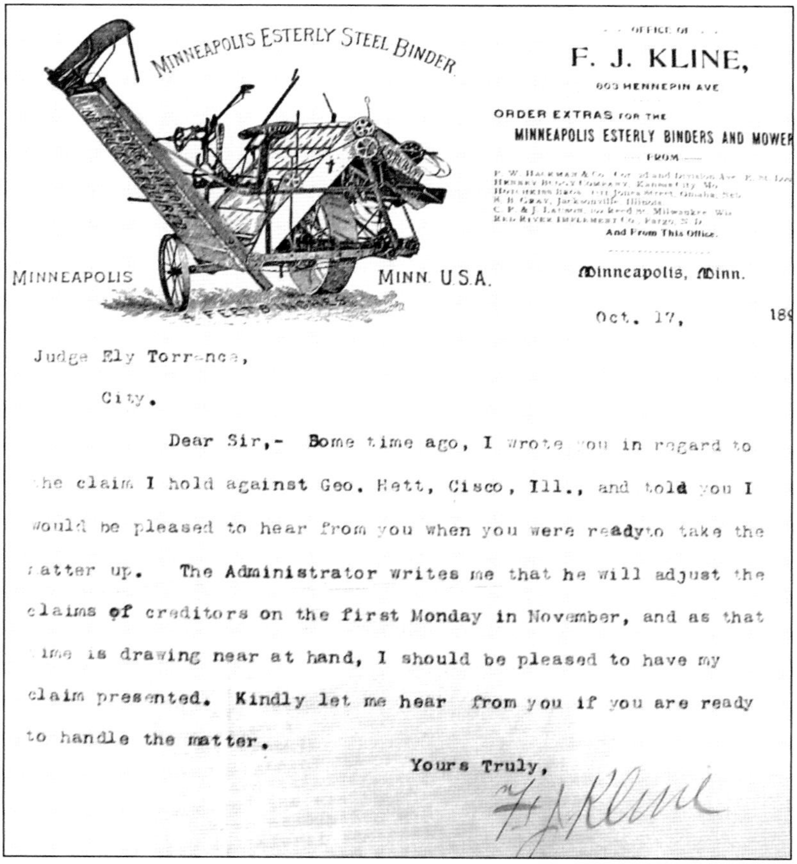

OFFICE OF

F. J. KLINE,

603 HENNEPIN AVE

ORDER EXTRAS FOR THE MINNEAPOLIS ESTERLY BINDERS AND MOWERS

FROM

And From This Office.

Minneapolis, Minn.

Oct. 17, 18

Judge Ely Torrance,

City.

Dear Sir,- Some time ago, I wrote you in regard to the claim I hold against Geo. Hett, Cisco, Ill., and told you I would be pleased to hear from you when you were ready to take the matter up. The Administrator writes me that he will adjust the claims of creditors on the first Monday in November, and as that time is drawing near at hand, I should be pleased to have my claim presented. Kindly let me hear from you if you are ready to handle the matter.

Yours Truly,

F J Kline

December 1894 saw the struggling company issue a deed of assignment, as its assets were redistributed among the many creditors clamoring for its money. By January, the company was insolvent, unable to pay its debts. In May 1895, Judge Eli Torrance ordered the foreclosure of the trust deed on the Esterly Company, at a value of $167,000. Pictured is a letter to Torrance from Esterly agent Frank Kline. (Minnesota Historical Society.)

This photograph shows an Esterly seeder plate, patented in 1868. This seeder plate was only one of the many products patented by Esterly during the 19th century. Seeder plates were used to ensure the regular and even distribution of seeds. This seeder plate was most likely used to seed corn and other large seeds. (Irvin L. Young Memorial Library.)

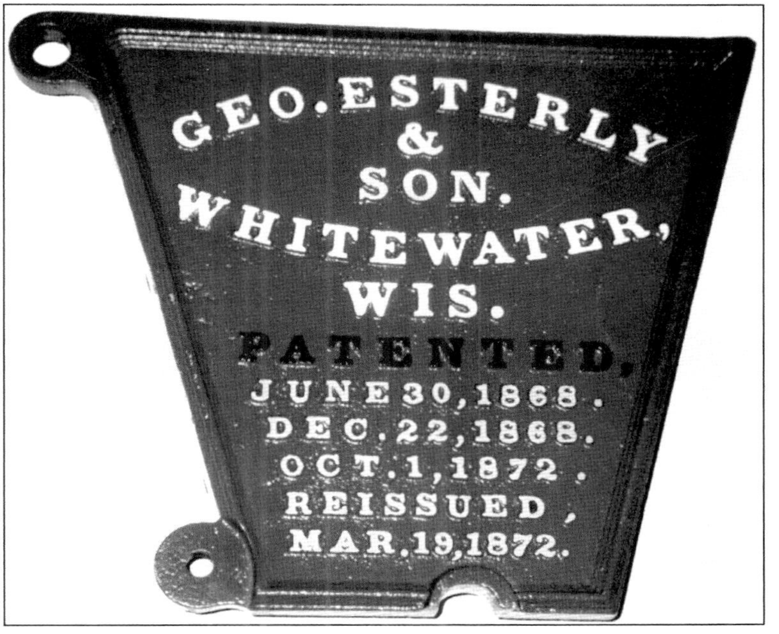

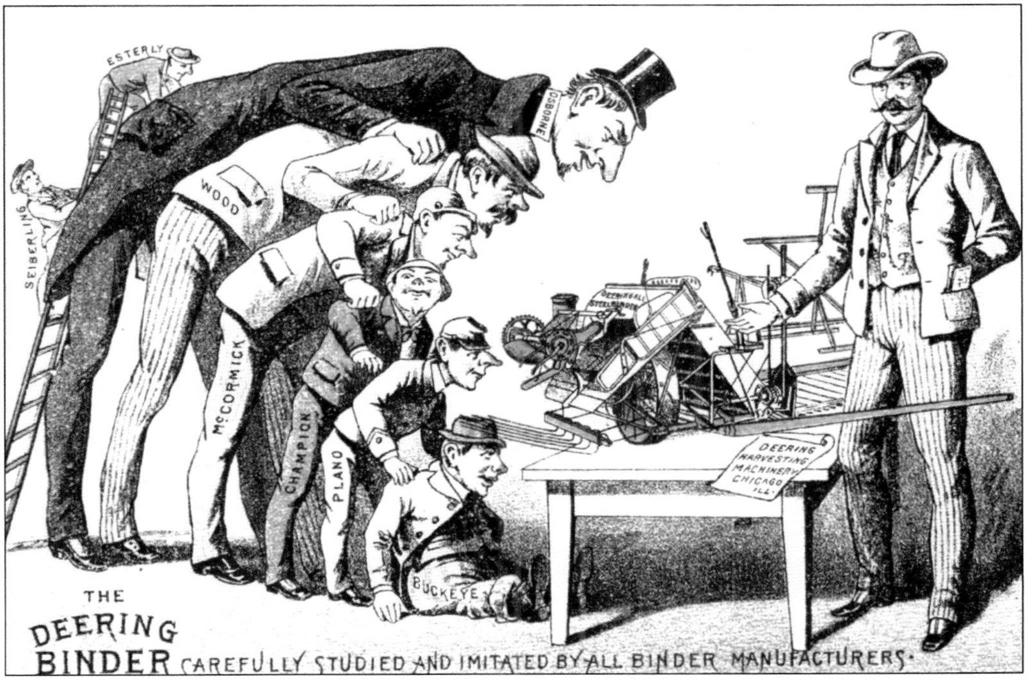

This advertising card from William Deering and Company of Chicago depicts Deering's chief rivals. George Esterly can be seen on the far left at the top of the ladder. This postcard shows that Esterly was known around the country as one of the most successful reaper men of his time. (Irvin L. Young Memorial Library.)

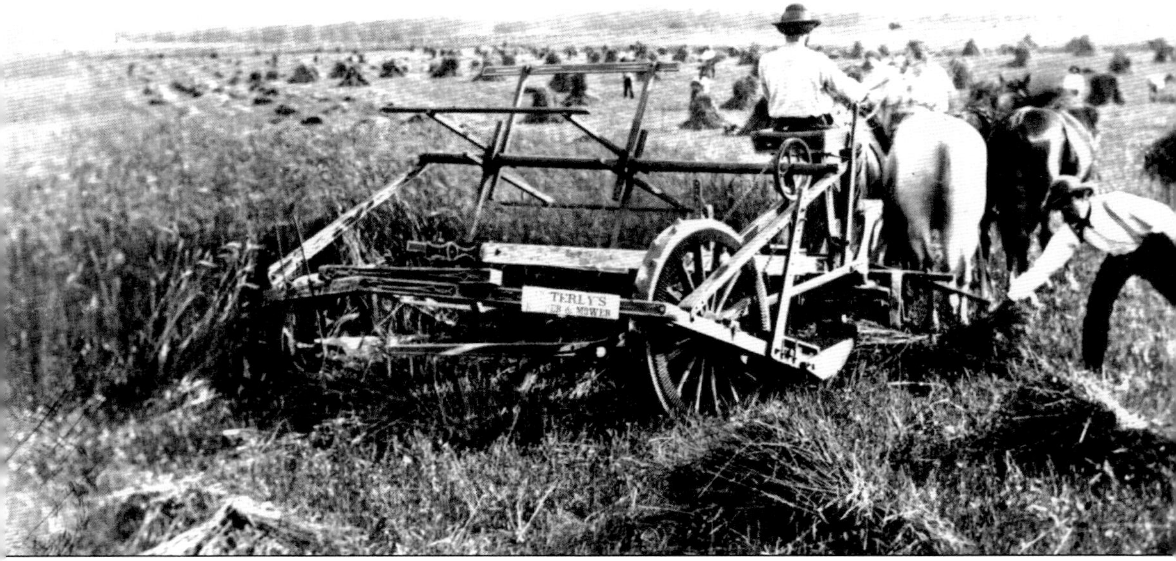

There is no single factor that completely explains the shutdown of the Minneapolis Esterly Company, as it was one of many thousands of businesses that failed in the wake of the panic of 1893. One can reasonably infer plummeting wheat prices were a major factor. The prairie farmer saw the average price of wheat drop more than 30 percent, naturally making it difficult for these farmers to continue to purchase agricultural implements. The Minneapolis Esterly Company was ultimately another casualty of a widespread economic disaster. Walworth County's largest business, one that had thrived in Whitewater only a few years before, would not survive in Minneapolis in the midst of a great national financial crisis. This c. 1860 photograph shows an Esterly Reaper in action in St. Paul, Minnesota, the only known photograph of the Esterly reaper from this time. (Minnesota Historical Society, Whitney's Gallery.)

Four

Winchester, Partridge, and the Whitewater Wagon

While the Esterly Manufacturing Company was Whitewater's most notable firm, another large manufacturer made Whitewater its home. The Winchester and Partridge Wagon Factory employed hundreds of men and enjoyed brisk sales, both nationally and internationally. The company also developed its own unique farm wagon that was produced for decades after the closing of the firm. For nearly 40 years before its closure, the Winchester and Partridge Manufactory helped sustain Whitewater and build its reputation as a manufacturing center. This company offers another example of businessmen, like Esterly, who achieved success in a relatively small market due to the perceived superiority of their product.

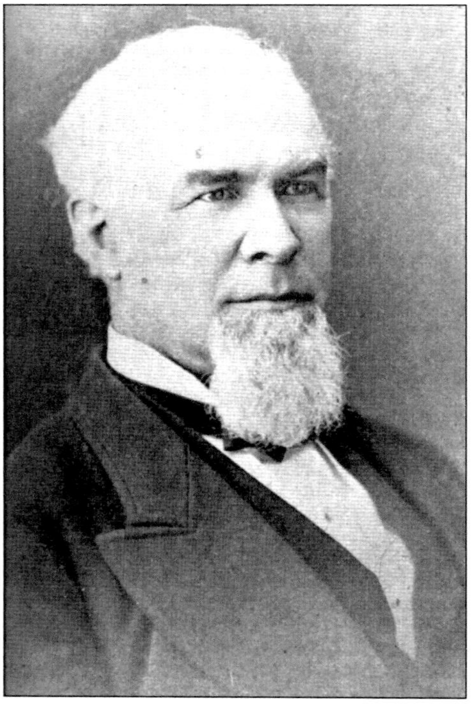

Lucius Winchester was born in Hartford, Vermont, in 1821. He came to Whitewater in March 1844 and went into the general blacksmithing business with Joseph Rogers, operating a shop until 1853. Winchester was a blacksmith by trade, but he attempted to break into plow manufacturing in 1850. He attached a foundry to his blacksmith shop, where he built his first one-horse plow. (WHS.)

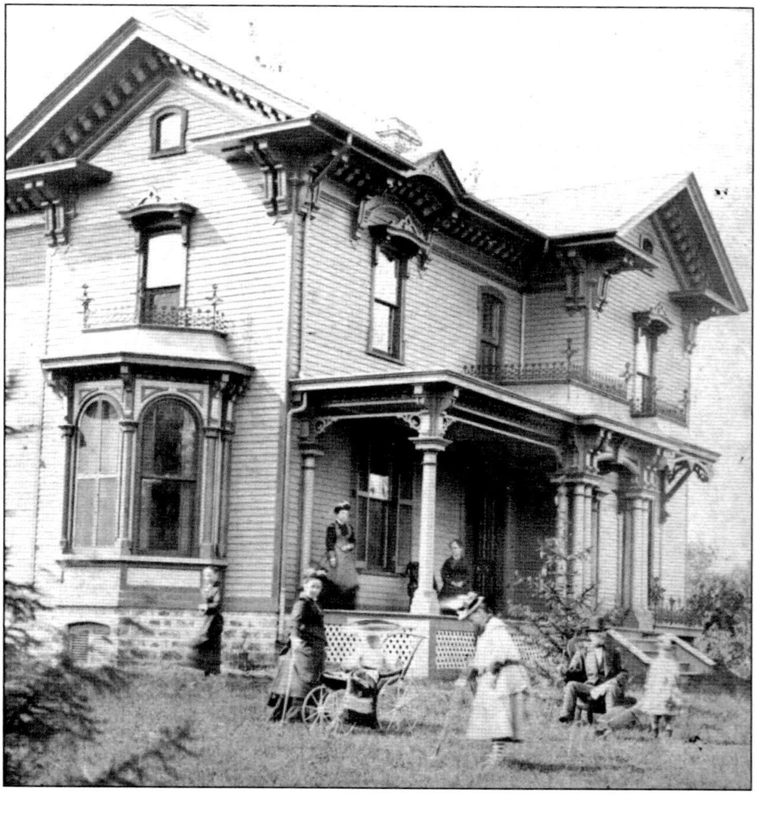

In 1853, William DeWolf moved to Whitewater and entered into a partnership with Lucius Winchester. They manufactured plows, initially employing seven men and one horse. DeWolf was a native New Yorker who had arrived in Wisconsin in 1842 with 75¢ to his name. This photograph shows the DeWolf home, built on the corner of Main Street and Prairie Street in 1873. (WHS.)

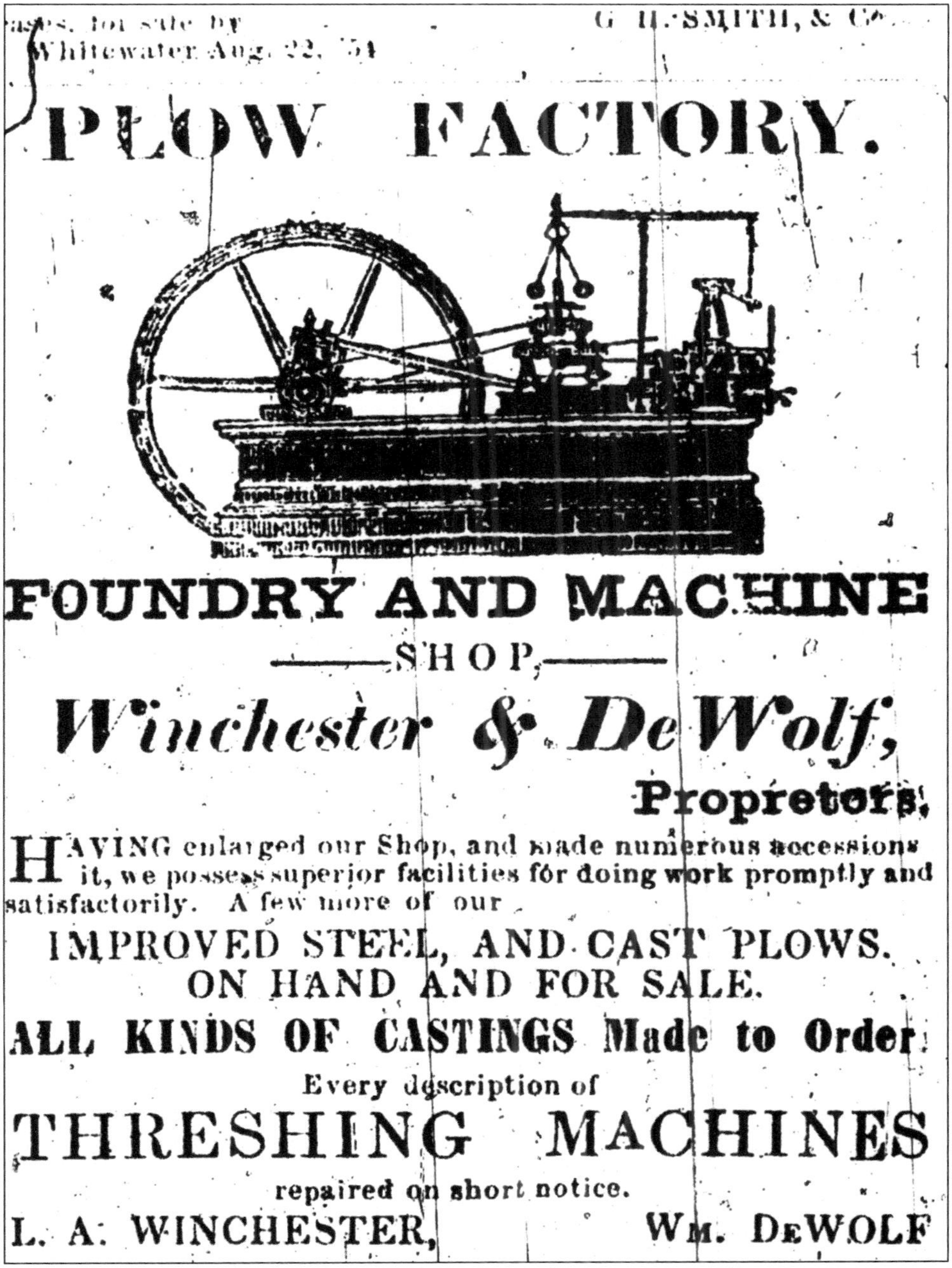

The Winchester and DeWolf Foundry and Machine Shop began as a plow manufactory with a foundry to produce castings. The men had constructed a plow factory in 1855 and soon began production. This advertisement in the first issue of the *Whitewater Gazette*, dated January 10, 1856, offered "all kinds of castings made to order" as well as "every description of threshing machines repaired on short notice." However, plows became the cornerstone on which the company was built. In fact, by 1857 the Winchester and DeWolf plow factory was the largest in the state. With more than 30 employees, the factory turned out more than 3,000 plows annually, which sold "almost wholly." (Whitewater Gazette.)

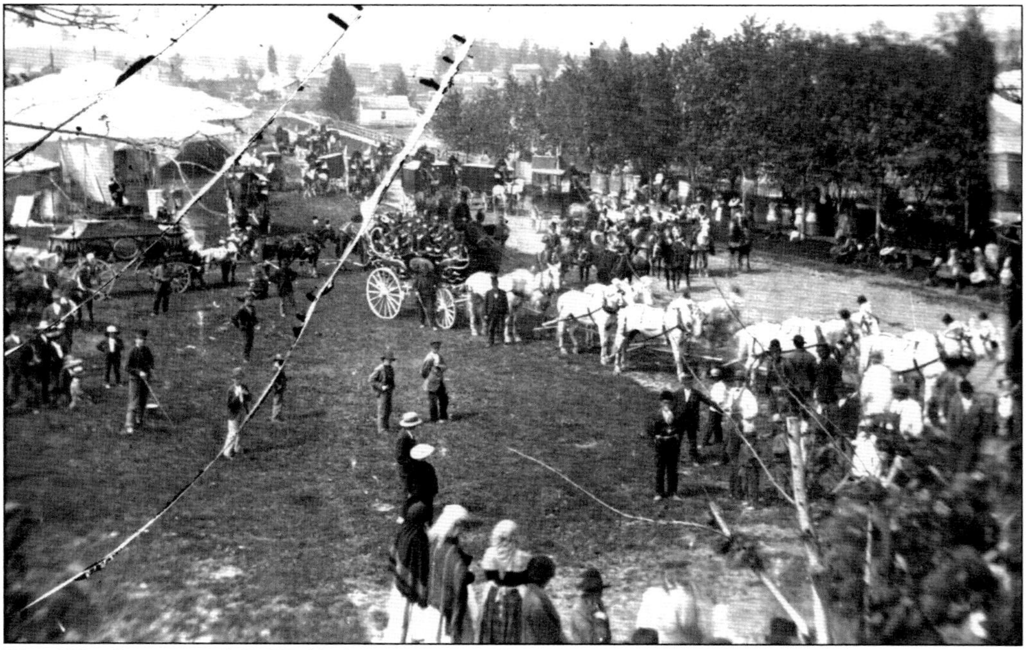

The Winchester and DeWolf plow was soon known around the state. Judges at the Jefferson County Fair awarded Winchester and DeWolf "best lot of ploughs" in October 1856. In the same month, the plow won the highest premium at the Dane County Fair (pictured), already "taking the highest prices all over the State, for [it was] the best." The plow also won multiple prizes at the Wisconsin State Fair. (Wisconsin Historical Society, image No. 27141.)

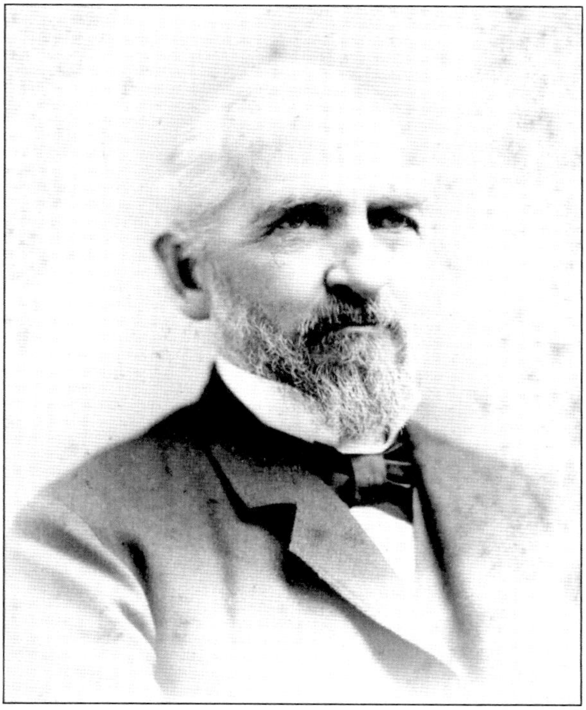

A New York native who came to Whitewater in 1845, John S. Partridge started a mercantile business along with Sanger Marsh, eventual organizer of the First National Bank of Whitewater. Partridge and Marsh built what was known as "the Big Warehouse" after the arrival of the Milwaukee and Mississippi Railroad in the early 1850s. In 1857, Partridge purchased a share of the Winchester and DeWolf Company. (WHS.)

John S. Partridge helped to organize the Farmers' and Merchants' Bank of Whitewater. He became a cashier for the bank in 1856. The Merchants' Bank was located on Center Street next to the Metropolitan Hall. The bank bought and sold foreign currency but issued no bills of its own. He built his home, pictured, on Center Street in 1856. (WHS.)

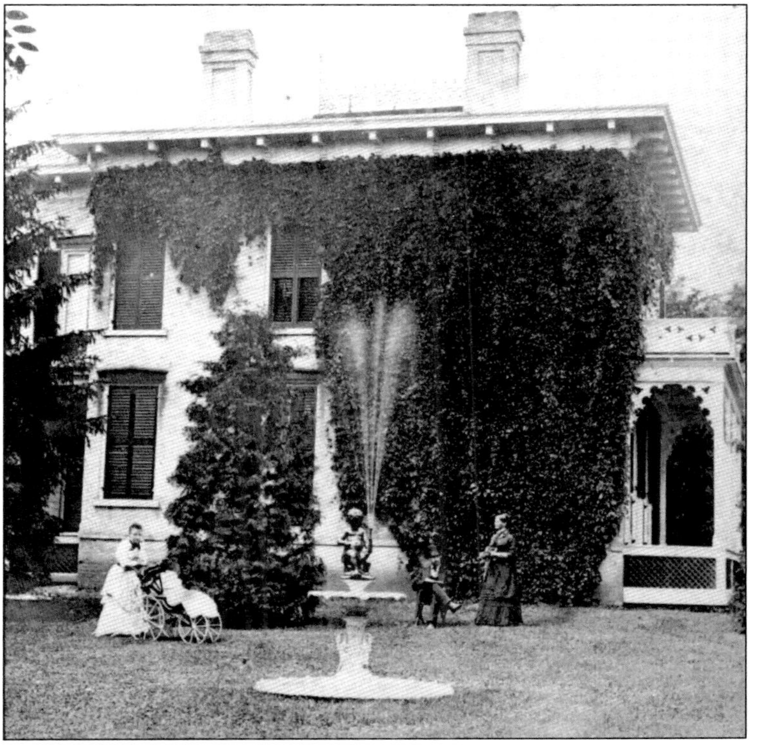

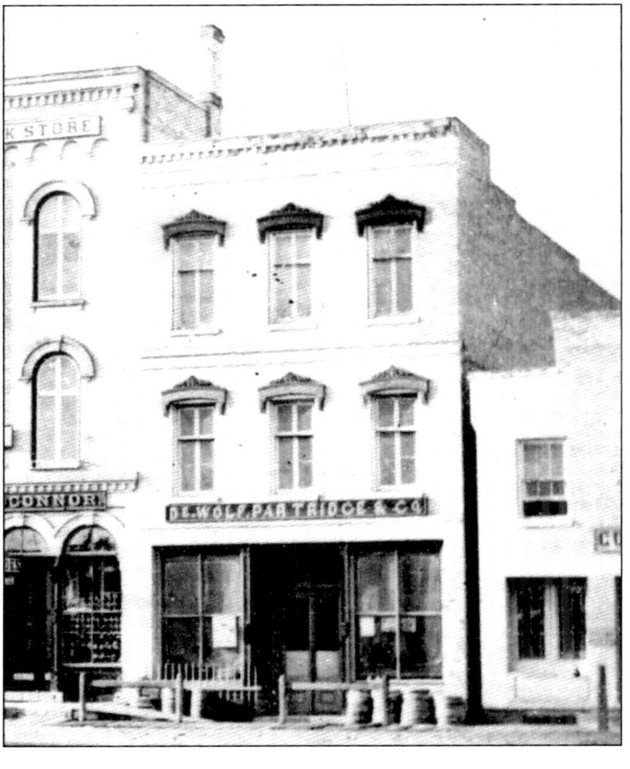

John S. Partridge closed out his other business ventures before purchasing his interest in Winchester and DeWolf, which then became Winchester, DeWolf and Company. In 1859, Winchester, DeWolf and Company expanded with the addition of a general store (pictured) run by Partridge, but its goods were sold less than a year later to George W. Cannon, so that exclusive attention could be paid to the manufacture of plows and other implements. (Irvin L. Young Memorial Library; Fran Achen Collection.)

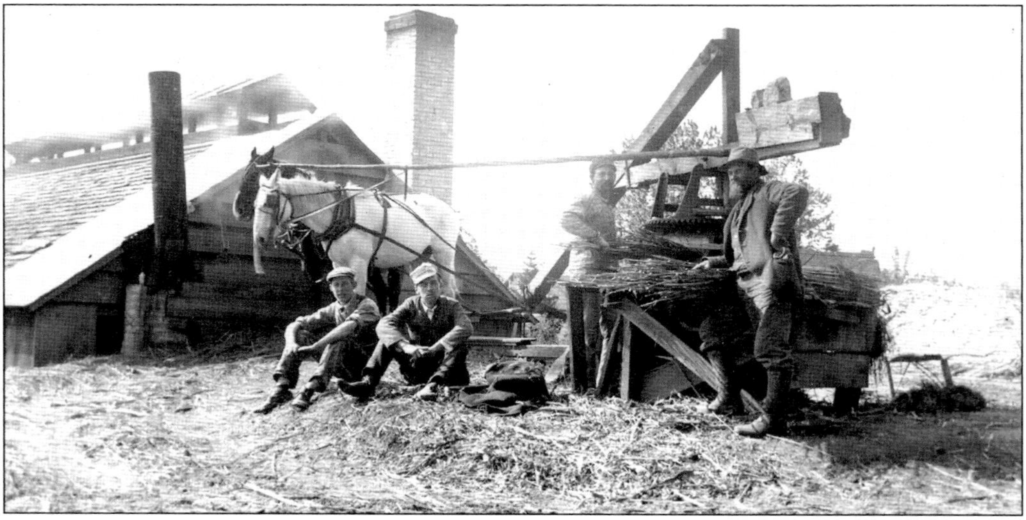

Bassett's Challenge Sugar Cane Mill was patented by Daniel Bassett and manufactured by Winchester, DeWolf and Company. This well-regarded mill produced sorghum syrup, which was a sweet substitute for sugar when processed sugar was not readily available. Observers of the mill happily noted that it left sorghum stalks "perfectly dry," and this mill was another early success for Winchester and DeWolf. This photograph shows workers preparing to process sorghum. (Irvin L. Young Memorial Library; Fran Achen Collection.)

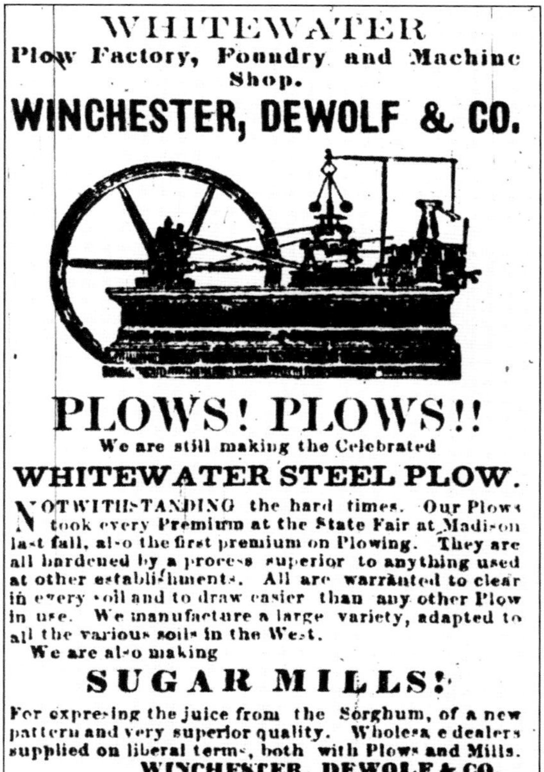

This 1859 advertisement shows a commonly used drawing of Bassett's mill. Bassett, a machinist employed by Winchester, DeWolf and Company, originally invented this machine for planting plow beams. The mill had tongued and channeled rollers. These rollers allowed the juice from the sorghum to be expressed in a continuous stream. Cane waste did not come in contact with the juice, keeping it pure. (Whitewater Register.)

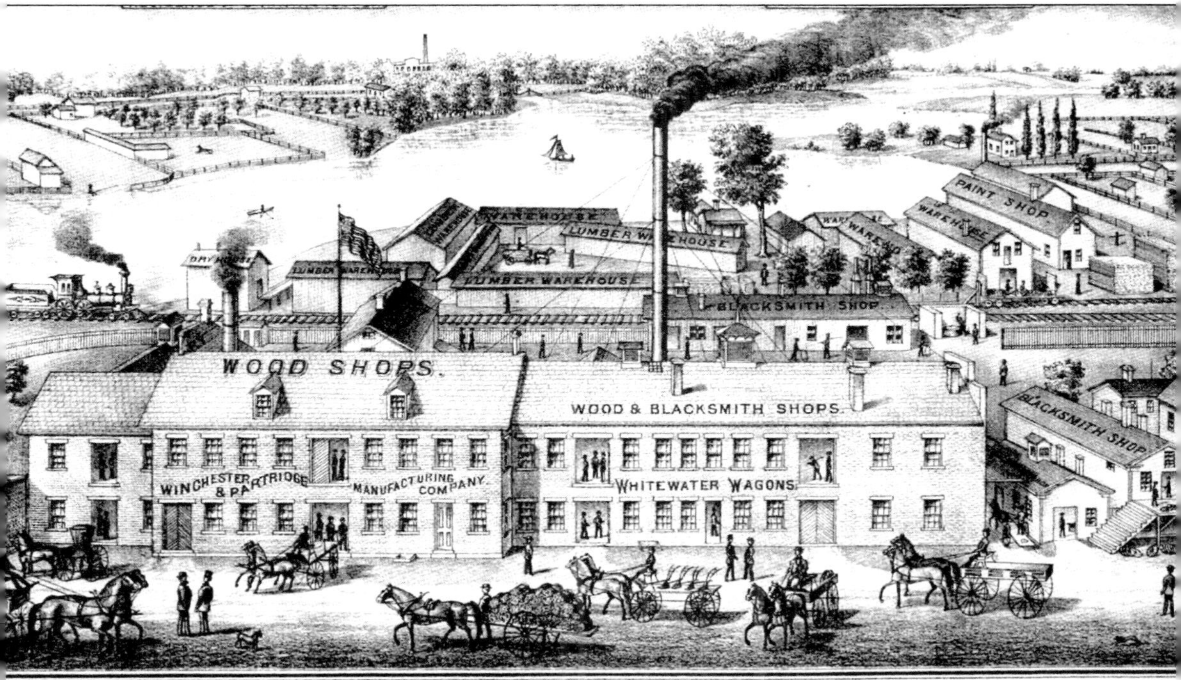

WHITEWATER WAGON & PLOW MANUFACTORY.
WINCHESTER & PARTRIDGE.

In the early morning hours of September 27, 1859, Winchester, DeWolf and Company faced its first serious setback. A destructive fire consumed the contents of its warehouse, destroying 800 plows, $4,000 to $5,000 worth of lumber, three mills, and various other items, with total damages of more than $14,000. The losses incurred were significant, and the delivery of plow orders was delayed for many weeks. Despite the delays, the company carried on manufacturing almost immediately. No concerns about relocation plagued the aftermath of the fire, as they later would following the great fire at the Esterly Works. By November, a new warehouse was ready for occupation and production resumed, with 3,800 steel and 1,000 cast-iron plows completed during 1859, in addition to 400 road scrapers, 500 breaking plows, and many sugar mills. (Zaballos family.)

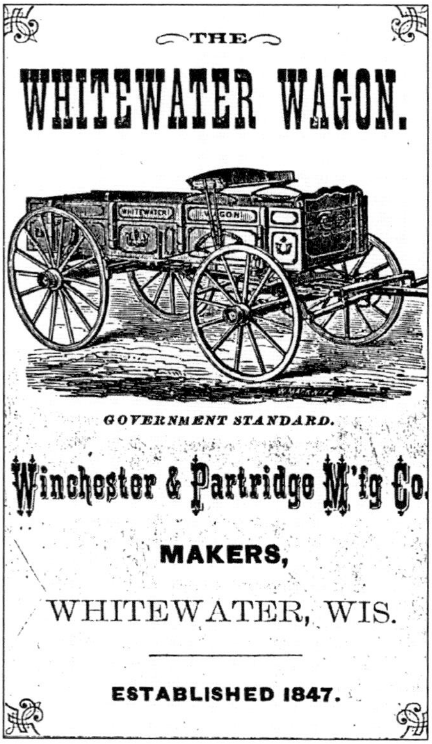

Winchester, DeWolf and Company's plow brought the business into regional prominence. But it was another product, a new farm wagon, which allowed the company to burst onto the national stage. The Whitewater Wagon, as it was eventually known, came into production around 1860. In September 1861, the *Whitewater Register* reported that Winchester, DeWolf and Company was producing 36 wagons a week, yet was still unable to fill consumers' demands. (Minnesota Historical Society.)

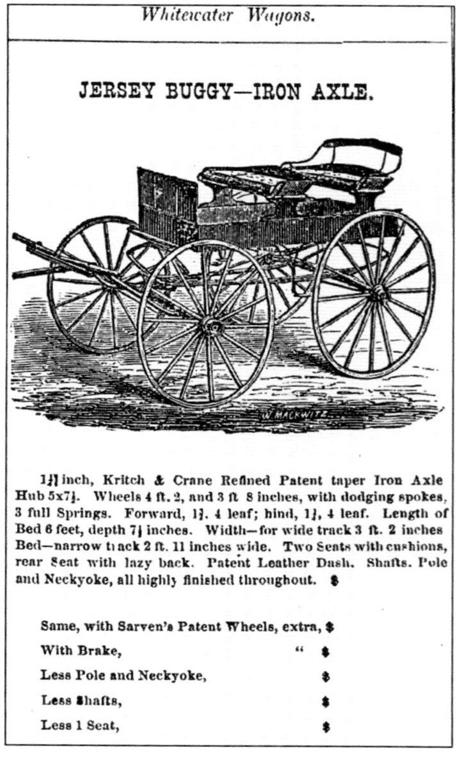

Whitewater Wagons.

JERSEY BUGGY—IRON AXLE.

1¼ inch, Kritch & Crane Refined Patent taper Iron Axle Hub 5x7½. Wheels 4 ft. 2, and 3 ft 8 inches, with dodging spokes, 3 full Springs. Forward, 1¾. 4 leaf; hind, 1¼, 4 leaf. Length of Bed 6 feet, depth 7½ inches. Width—for wide track 3 ft. 2 inches Bed—narrow track 2 ft. 11 inches wide. Two Seats with cushions, rear Seat with lazy back. Patent Leather Dash. Shafts. Pole and Neckyoke, all highly finished throughout. $

Same, with Sarven's Patent Wheels, extra, $
With Brake, " $
Less Pole and Neckyoke, $
Less Shafts, $
Less 1 Seat, $

A three-story brick building was soon constructed to produce more wagons, and the Whitewater Wagon's fame began to spread. Wagon production became a priority in 1865, with the company advertising its lumber wagons for sale in any quantity. At the end of 1865, wagons were sent to St. Louis, with the expectation that they would be "known all over the western country." (Minnesota Historical Society.)

The Whitewater Wagon sold for $110 at Semple, Birge and Company on St. Louis's Main Street. Semple, Birge and Company exhibited the Whitewater Farm Wagon at the 12th annual fair of the St. Louis Agricultural and Mechanical Association in 1872, by which point *Prairie Farmer* magazine characterized the wagon as "celebrated through the West for its excellence." (Minnesota Historical Society.)

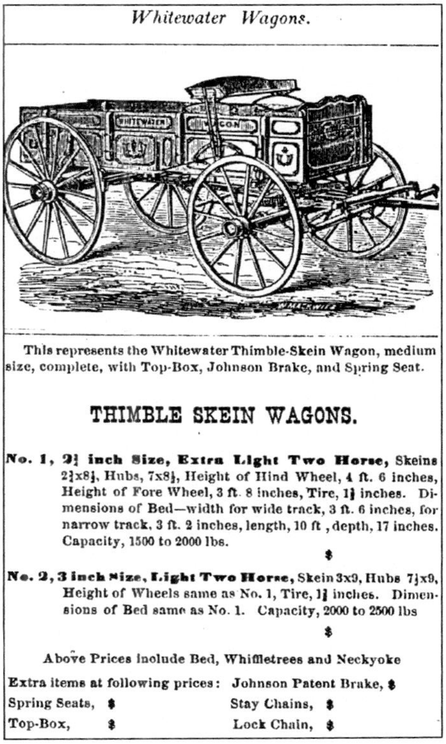

Whitewater Wagons.

This represents the Whitewater Thimble-Skein Wagon, medium size, complete, with Top-Box, Johnson Brake, and Spring Seat.

THIMBLE SKEIN WAGONS.

No. 1, 2¾ inch Size, Extra Light Two Horse, Skeins 2¾x8½, Hubs, 7x8½, Height of Hind Wheel, 4 ft. 6 inches, Height of Fore Wheel, 3 ft. 8 inches, Tire, 1½ inches. Dimensions of Bed—width for wide track, 3 ft. 6 inches, for narrow track, 3 ft. 2 inches, length, 10 ft, depth, 17 inches. Capacity, 1500 to 2000 lbs.

$

No. 2, 3 inch Size, Light Two Horse, Skein 3x9, Hubs 7¼x9, Height of Wheels same as No. 1, Tire, 1¾ inches. Dimensions of Bed same as No. 1. Capacity, 2000 to 2500 lbs

$

Above Prices include Bed, Whiffletrees and Neckyoke

Extra items at following prices: Johnson Patent Brake, $

Spring Seats, $ Stay Chains, $

Top-Box, $ Lock Chain, $

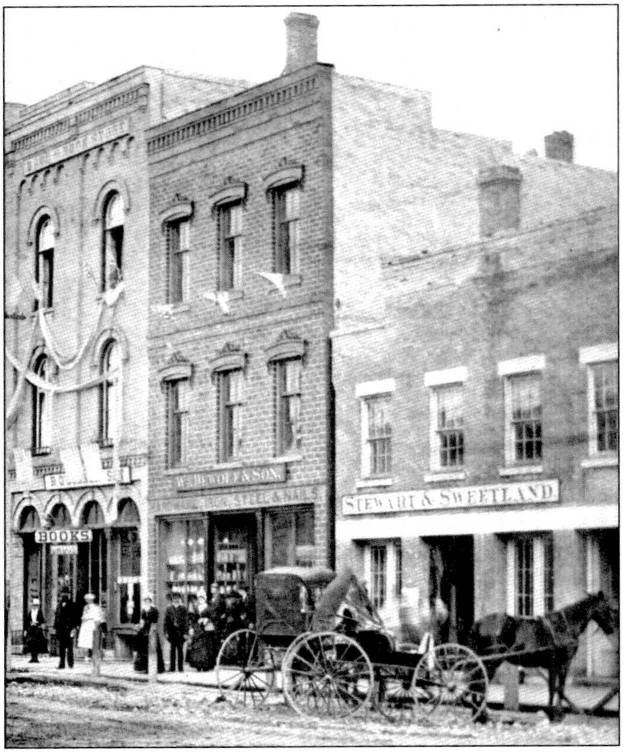

On January 7, 1867, Winchester, DeWolf and Company came to an end. DeWolf took possession of the company-owned hardware business and built a new store on Center Street, while Winchester and Partridge assumed the manufacture of wagons as Winchester, Partridge and Company. The DeWolf Wire Cloth Manufacturing Company (pictured) was founded in 1877 by William DeWolf and his son Mannering. Wire cloth was used for screens for homes and businesses. (Irvin L. Young Memorial Library; Fran Achen Collection.)

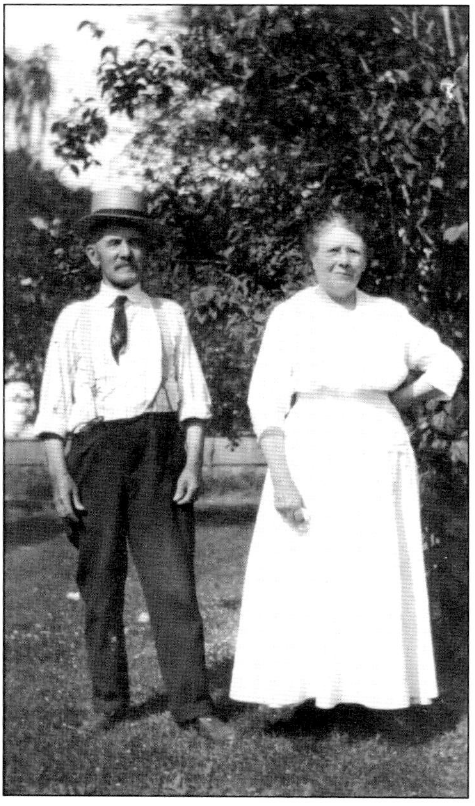

This photograph shows Mannering DeWolf with his wife, Effie Salisbury, in July 1918. Effie was the daughter of Ansel Salisbury, one-time owner of Whitewater's Branch Mill. Mannering was an officer of the Custer Rifles, a military unit that was first organized in Whitewater in 1877. Mannering worked as a milkman into his 60s. (WHS.)

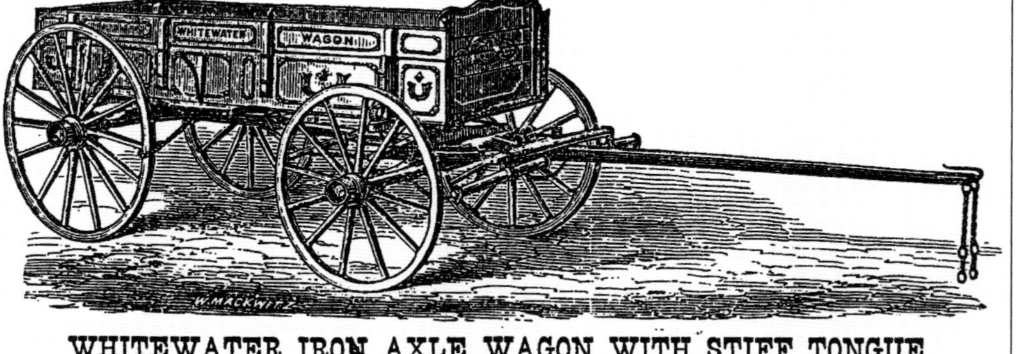

WHITEWATER IRON AXLE WAGON WITH STIFF TONGUE.

Winchester and Partridge produced a very successful wagon during the 1860s, and as such, they were soon to attract more notable and distant clients. In 1873, wagon orders came in from California to Connecticut, as well as the West Indies and the Sandwich Islands. The Whitewater Wagon also came into general use on South American coffee plantations. This international fame brought new attention to Whitewater. (Minnesota Historical Society.)

Winchester and Partridge's most famous client came from England, as Queen Victoria's 1873 North American Boundary Commission used the wagons for an expedition to determine northwestern boundaries between the United States and Canada. The Whitewater Wagon was selected for the expedition after "a thorough examination of all the different wagons made in the country." (Minnesota Historical Society.)

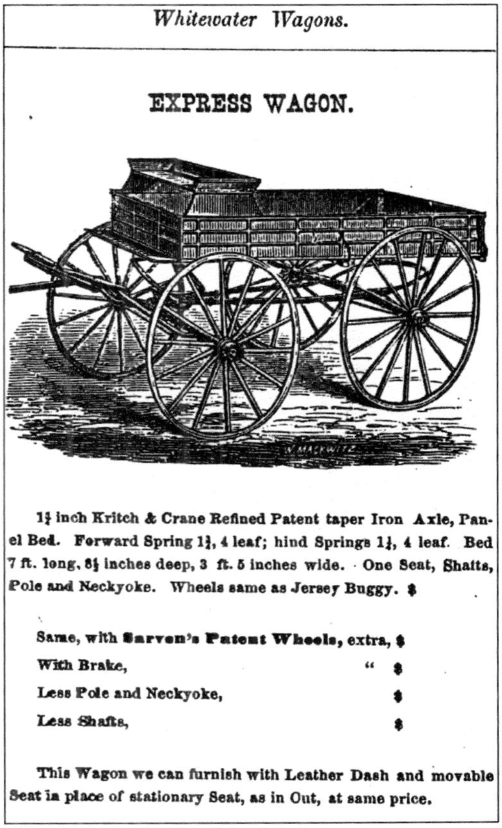

Whitewater Wagons.

EXPRESS WAGON.

1½ inch Kritch & Crane Refined Patent taper Iron Axle, Panel Bed. Forward Spring 1½, 4 leaf; hind Springs 1¼, 4 leaf. Bed 7 ft. long, 8½ inches deep, 3 ft. 5 inches wide. One Seat, Shafts, Pole and Neckyoke. Wheels same as Jersey Buggy. $

Same, with Sarven's Patent Wheels, extra, $
With Brake, " $
Less Pole and Neckyoke, $
Less Shafts, $

This Wagon we can furnish with Leather Dash and movable Seat in place of stationary Seat, as in Out, at same price.

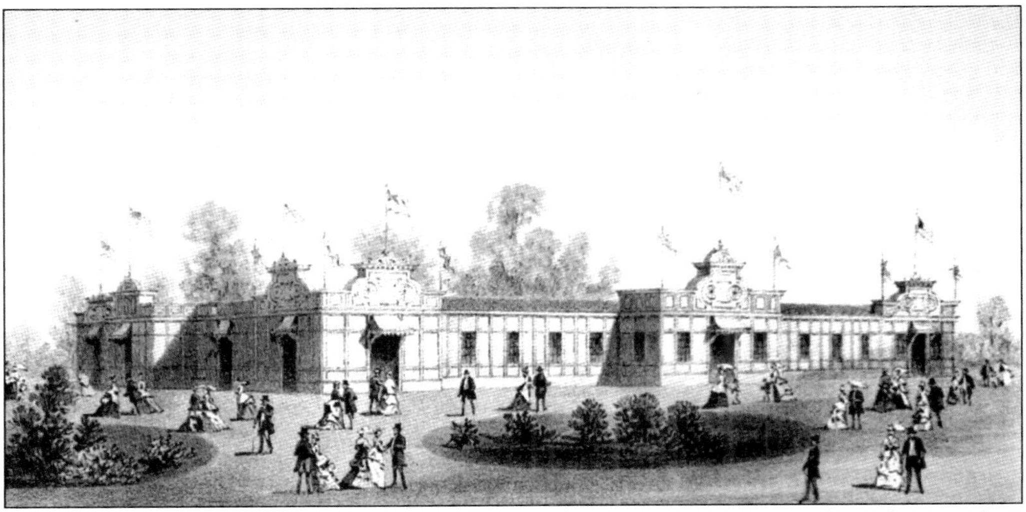

Wagon production increased in the 1870s to more than 3,000 wagons annually, with more than 150 men employed in the factory in 1873. In 1876, a great centennial exhibition was held in Philadelphia, and the Whitewater Wagon won a medal for excellence, as well as the "highest award of merit." This lithograph shows the carriage building at the exhibition where Winchester and Partridge exhibited three wagons. (Wikipedia.)

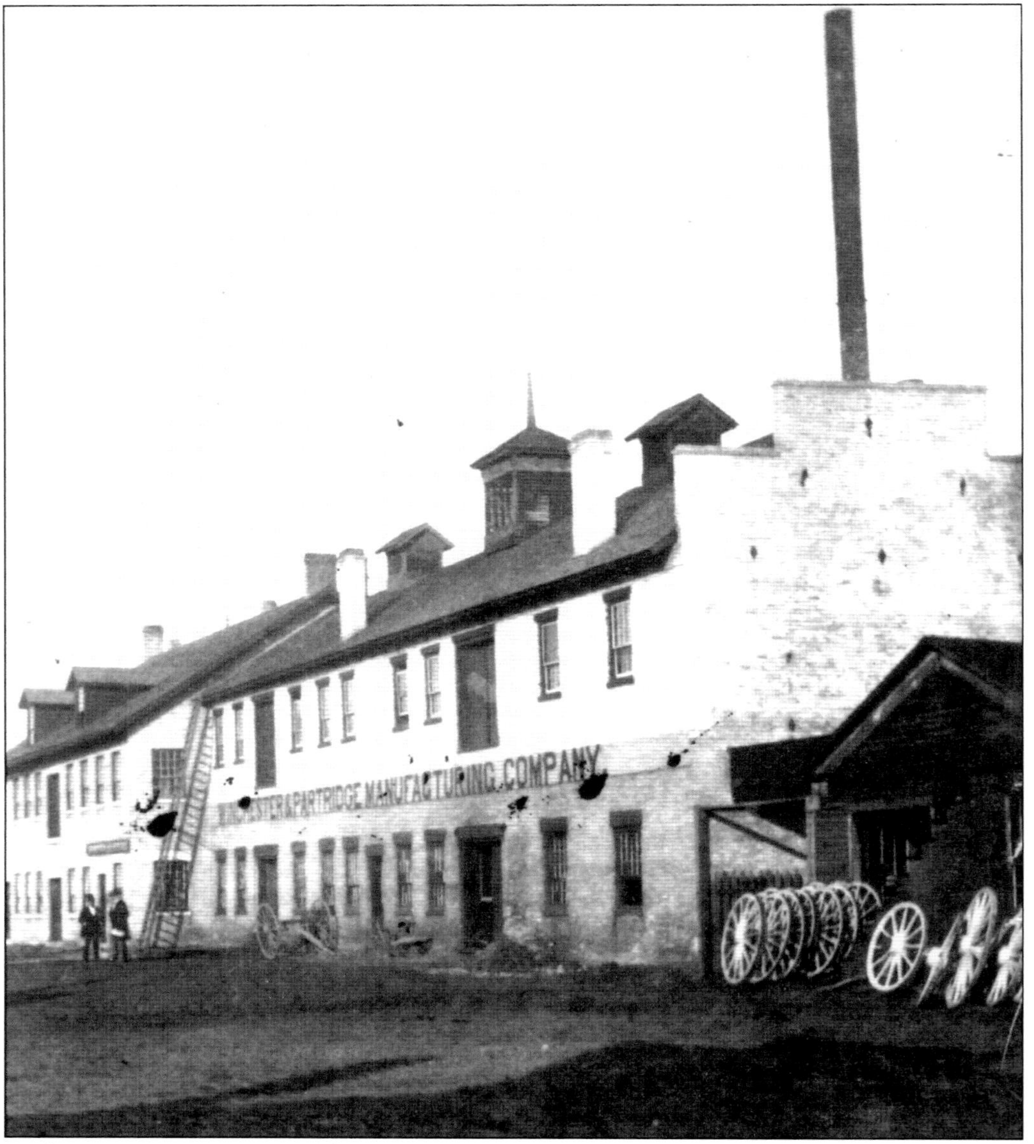

With new success came rumblings of a move to a larger city. In 1879, the *Milwaukee Sunday Telegraph* published a report that Winchester and Partridge planned to move to Milwaukee, and that negotiations were underway for the acquisition of 13 acres of land and buildings to be used as machine shops. A move to Milwaukee would allow Winchester and Partridge to expand employment to 300–400 men, as well as provide $200,000 in new capital. The *Whitewater Register* published an extensive response, stating that "[i]f the Wagon Factory should leave Whitewater there would be but little prospect left of its ever becoming a more important manufacturing town than it now is." Winchester and Partridge apparently paid more than $1 million in wages between 1864 and 1879, contributing greatly to the town's economy. The *Whitewater Register* implored the company to "sacrifice their personal advantage for the good of the town which has been their home." Ultimately, no such move occurred, and the company remained in Whitewater for what would be its last decade of operation. (WHS.)

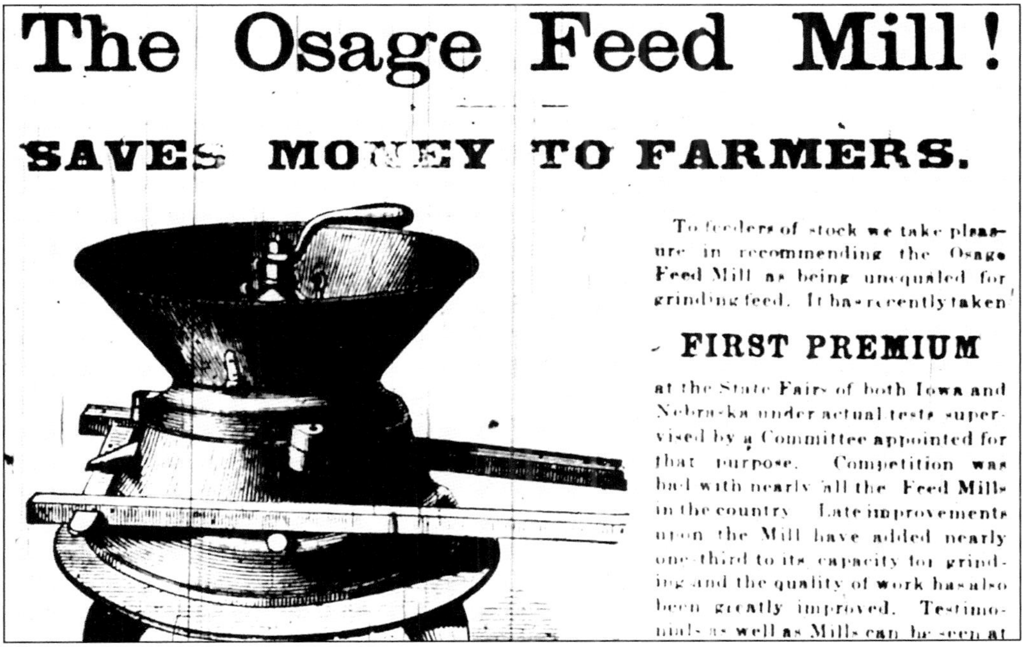

The 1880s brought further product diversification. During this decade, Winchester and Partridge began to manufacture the Osage Feed Mill, which combined a corn sheller with a feeder and won the first premium at both the Iowa and Nebraska state fairs. The Osage Feed Mill shelled the corn and ground it for feed. But the wagon remained at the forefront, with production peaking at more than 5,000 in 1882. (Whitewater Register.)

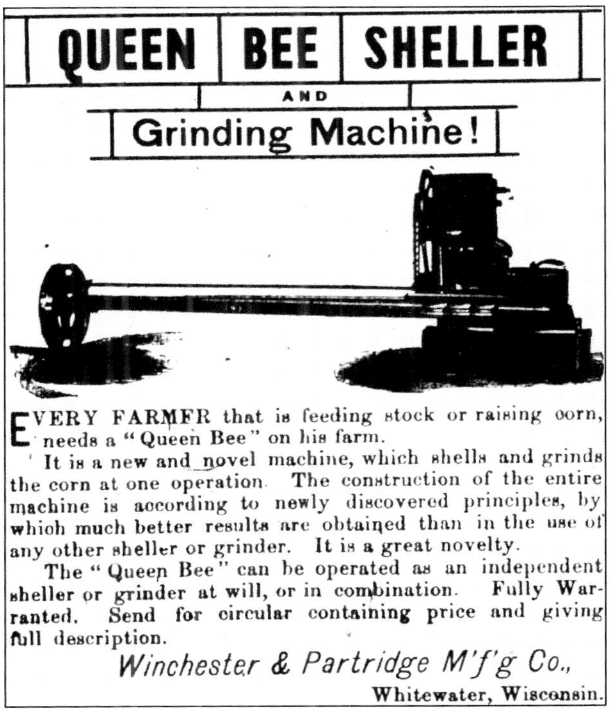

The Queen Bee, introduced in 1884, was a new type of corn sheller designed by E. M. Wilcox. This sheller separated corn from the cob and ground kernels to the desired fineness. Shellers greatly increased the efficiency of corn farming, as two men working together on a sheller could generate 60 to 80 bushels of grain corn in a single day. (Whitewater Register.)

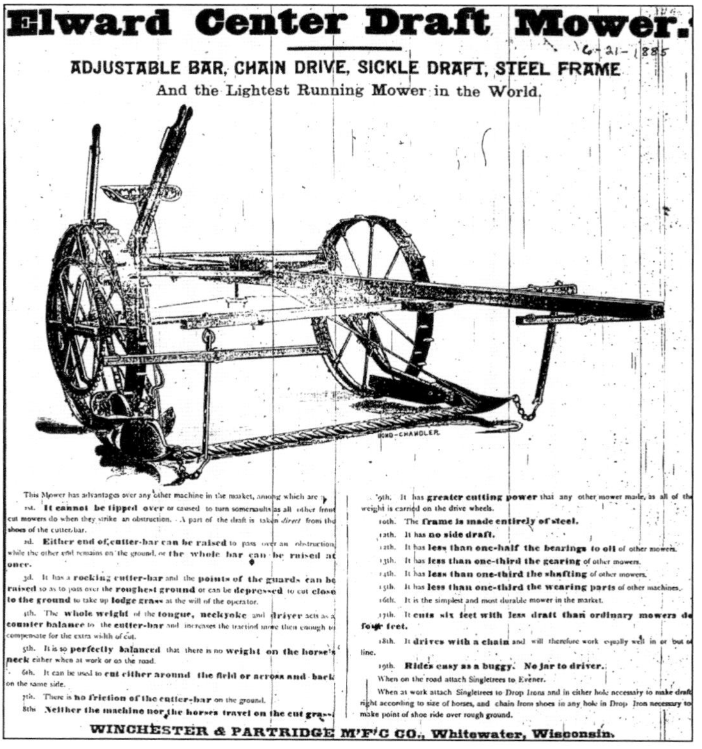

The Elward Center Draft mower, first introduced in 1885, was invented by Maj. John H. Elward, a Civil War veteran and lifelong inventor. In the early 1880s, Elward was the superintendent at the Minnesota Thresher Company in Stillwater, Minnesota. The Center Draft required a team of two horses to operate, one walking in the grass and another walking along the border. The mower was equipped with a six-foot sickle. (Whitewater Register.)

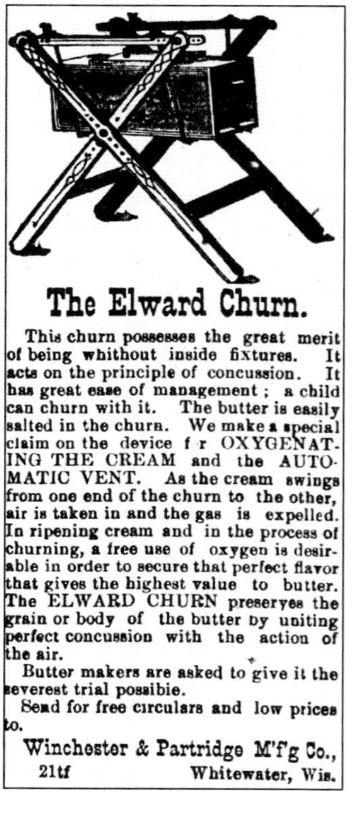

Butter and cheese were often produced on a small scale by local farmers who only made dairy products for their own use. The Elward Dairy Churn, first produced in 1886, was another of John Elward's inventions. This churn was endorsed by many of Wisconsin's leading dairymen. The increased efficiency of this churn allowed farmers to easily produce a surplus of butter that could be sold to local merchants. (Whitewater Register.)

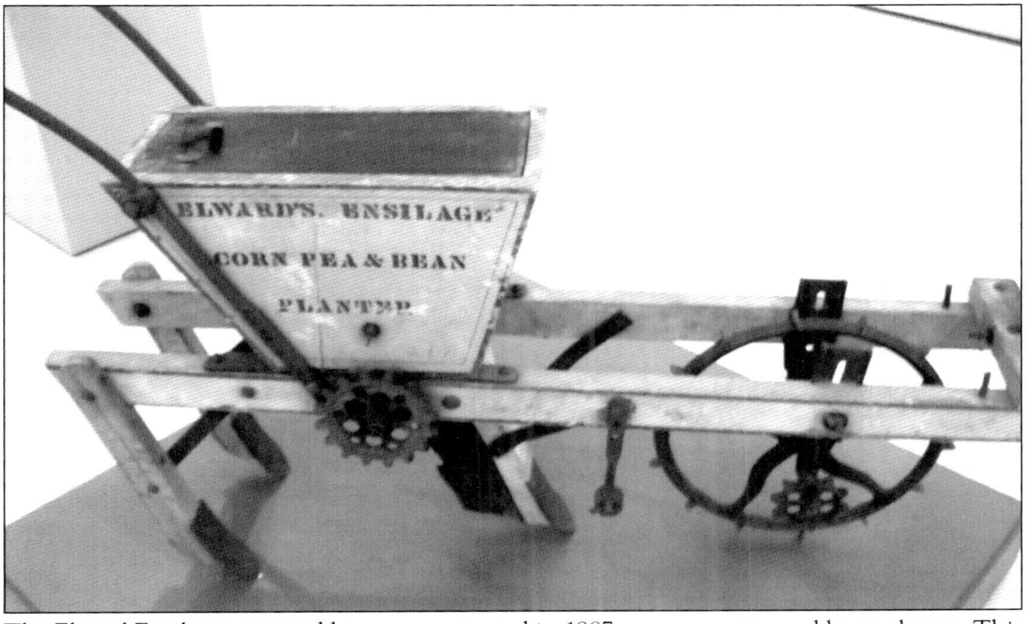

The Elward Ensilage, pictured here, was patented in 1887 as a corn, pea, and bean planter. This machine was an adaptation of an earlier cotton-planting model, developed by Elward in 1885. The machine both planted and covered seeds. The machine required only one horse to draw it, with one person guiding behind. This particular model can be seen at the Whitewater Historical Society Museum. (Author's collection.)

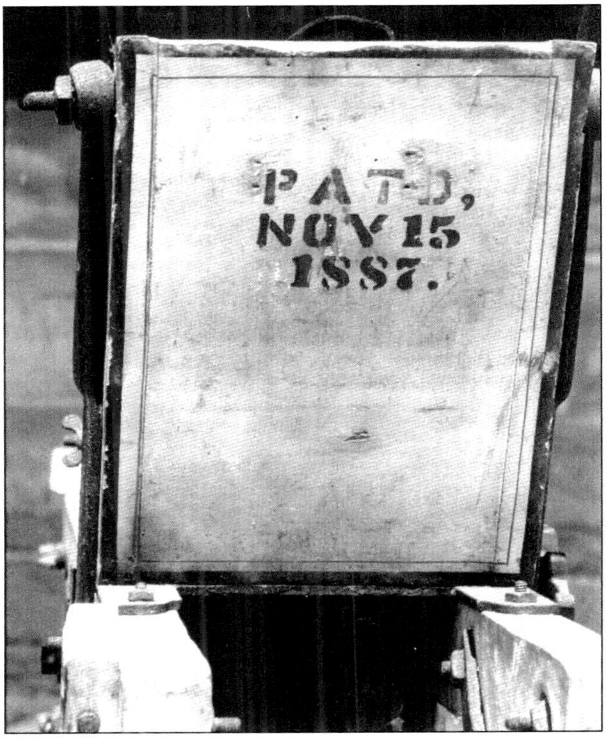

The back of this Elward Ensilage shows the date of Elward's patent, November 15, 1887. Orders for this planter poured in from all parts of the country, and the Winchester and Partridge Company could hardly produce enough planters to meet demand. By this point, the company had to employ more than 125 workers to fill orders for their wagons, churns, and cotton planters. (Irvin L. Young Memorial Library.)

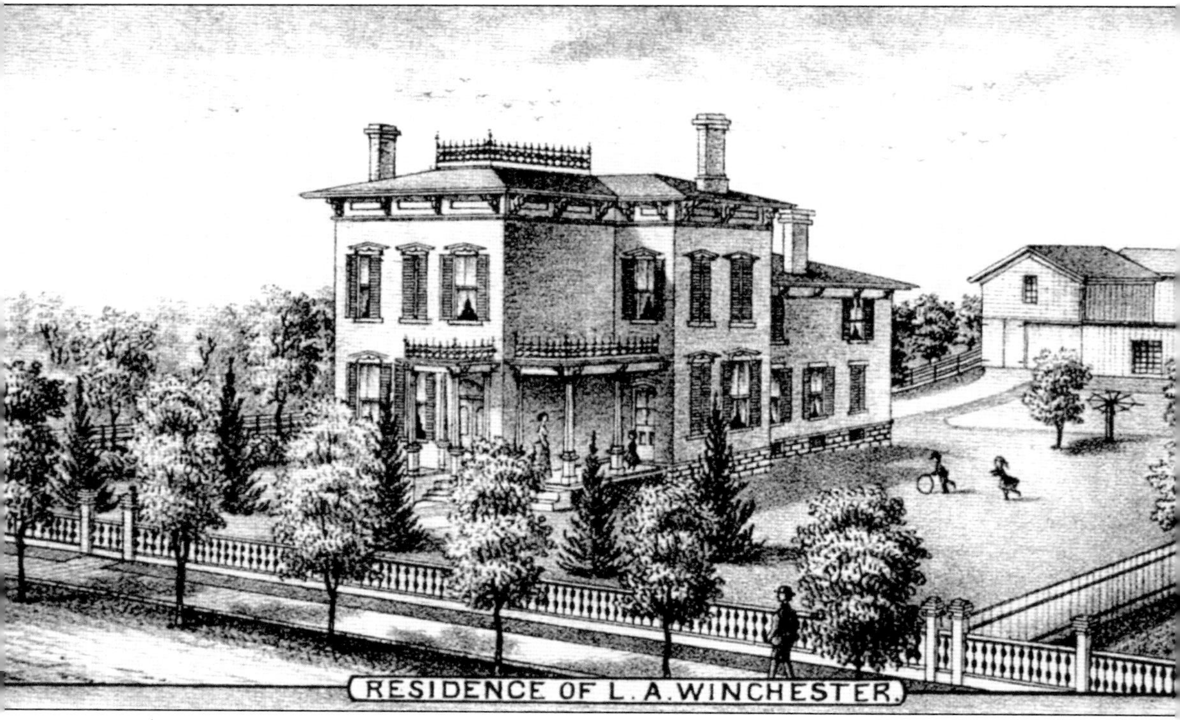

Both Winchester and Partridge were growing old by the late 1880s, and they had hoped to pass their factory on to a suitable heir. The Winchester sons were in line to inherit the family business, but a series of tragic events made this impossible. Frank Winchester died in 1881 at age 27 of typhoid pneumonia. The elder Winchester son had just graduated from college and was his father's likely successor. However, this responsibility now fell to his younger brother Charles. Charles Wakeley Winchester captured the imagination of the state in 1887 when he disappeared without a trace. He returned to Whitewater some months later with no real explanation, although many suspect that Winchester created his own disappearance to remedy the fact that he was engaged to two women at the same time. Charles, the last hope for the company, died suddenly at age 29 on August 18, 1889. After the second unexpected death, Winchester and Partridge decided to liquidate the company. The Winchester home, pictured, is located on Main Street. (Zaballos family.)

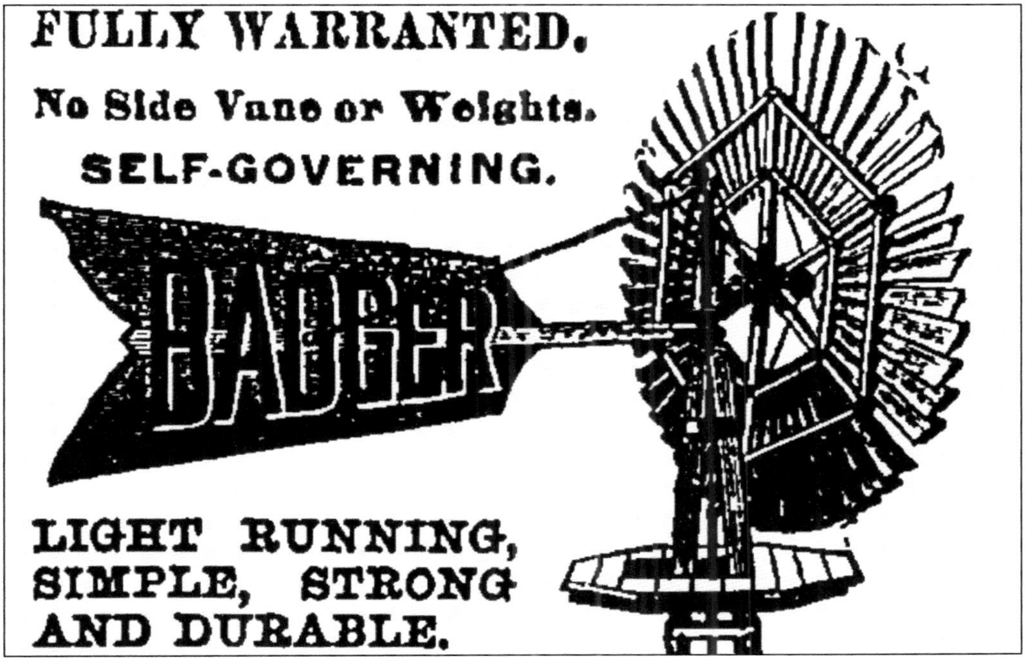

In 1889, Winchester and Partridge discontinued the manufacture of wagons and sold their entire stock. They closed out their stock of all remaining goods in early 1890 by selling everything at "cost prices." The company continued to operate in some limited capacity until at least 1891, as advertisements ran in *Southern Cultivator* magazine for a windmill manufactured by Winchester and Partridge, imploring readers to "send for our catalogue." (Southern Cultivator.)

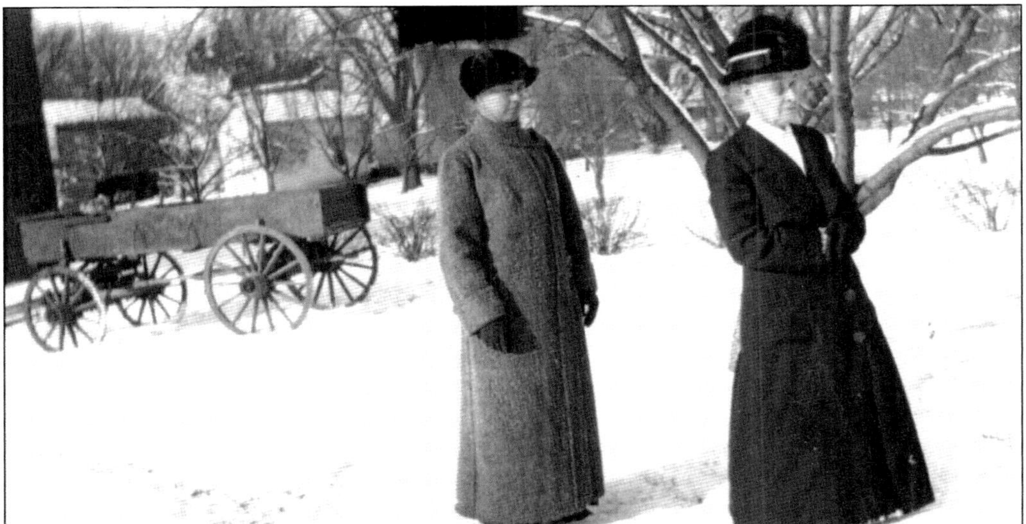

On April 9, 1890, Lucius Winchester died of Bright's disease. The next year, John Partridge retired from the bank due to his declining health. Partridge was a beloved member of the community whose "heart naturally went out to make a friend of every one." Partridge died on July 3, 1892. However, his wagons remained an integral part of life in Whitewater, as evidenced by this early-20th-century photograph. (WHS.)

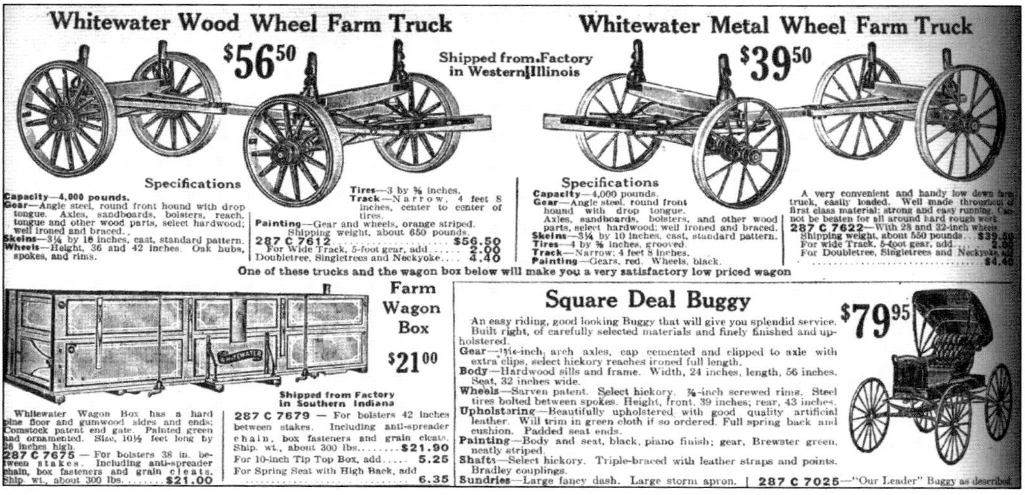

The legacy of Winchester, Partridge, and DeWolf lived on. The Whitewater Wagon name carried weight across the country, as Montgomery Ward and Company can be found selling lines of Whitewater Wagons as early as 1903 and as late as 1930. Montgomery Ward and Company appears to have purchased Winchester and Partridge's patents. This image is from the 1922 Montgomery Ward Catalog. (Wisconsin Historical Society.)

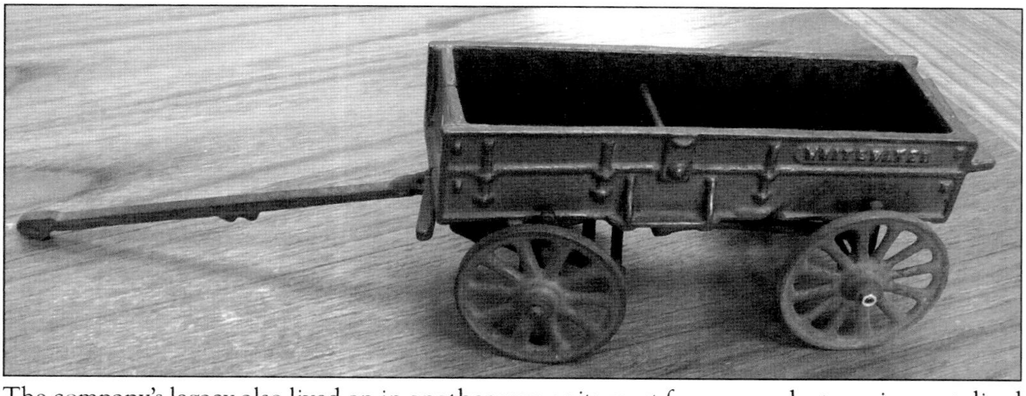

The company's legacy also lived on in another way, as its most famous product was immortalized as a children's toy. This Whitewater Wagon toy is a replica of Winchester and Partridge's basic Whitewater Wagon. This model was part of a line of cast-iron toys manufactured by National Sewing Machine Company of Belvidere, Illinois. (Coburn Company.)

Five

The Decline of an Industrial Town

In the 1880s, Whitewater was a thriving industrial town. Its farms and industries were growing, and the town's population increased at its fastest-ever rate. But by the dawn of the 20th century, the town was floundering, unable to maintain its population and economic productivity. Over the course of only two decades, one of Wisconsin's fastest-growing and most promising towns became a shell of its former self. Whitewater lost a quarter of its population and much of its notability.

In 1890, while George Esterly's company was still thriving in Whitewater, and Winchester and Partridge had only recently ceased operation, the town's population was 4,359. Just five years later, after Esterly's move, Whitewater claimed only 3,799 residents. The decline in population continued, with the city hitting a low point in 1905 with 3,108 residents. However, the county as a whole had grown by five percent from 1890 to 1900. The loss of the town's most important businesses significantly retarded Whitewater's growth, as no industries quickly emerged to replace them.

In the last years of the 19th century, it appears that Whitewater's citizens were busy blaming each other for the town's struggles without searching for a solution. An 1895 editorial in the *Whitewater Register* discusses the problems that developed from a "spirit of dissension, proscription and of unreasoning fault-finding [which] grows like a fungus when once started and fattens on its own poison. It is a dangerous and threatening element; its influence has already shown itself harmfully in our midst." A "spirit of criticism" descended over town, and the author lamented its negative impact on public improvements as well as public officials who grew increasingly weary of their public role in a contentious environment.

This chapter describes many of the smaller businesses and individuals who supported the town through its most difficult times. The later years of the 1800s were the beginning of a transition for Whitewater, and the necessary adjustments brought about the basis for recovery.

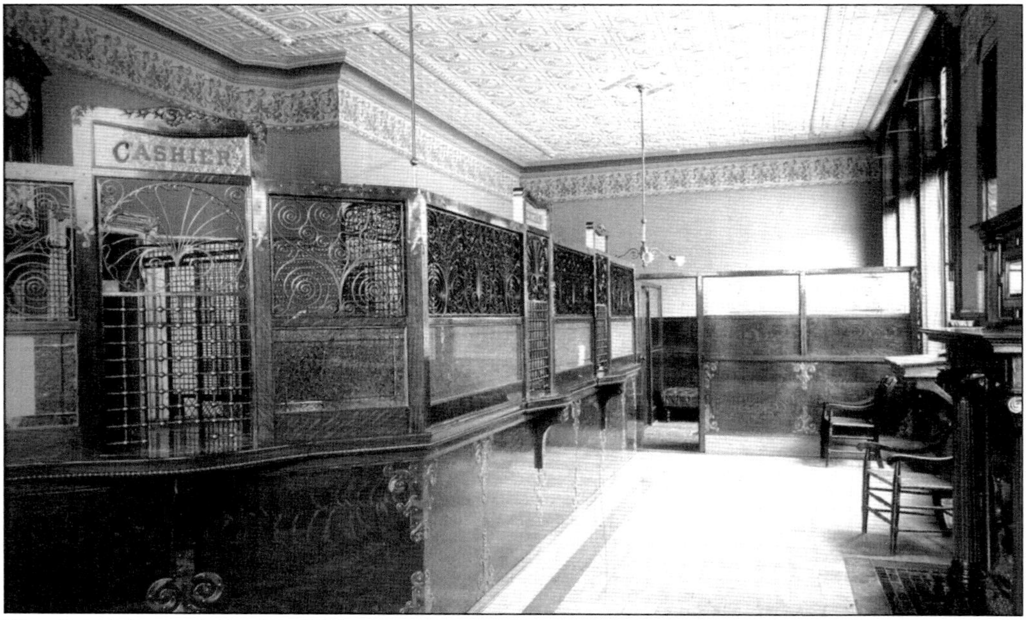

The First National Bank of Whitewater was chartered on October 25, 1863. The first officers of the bank were Pres. Sanger Marsh, Vice Pres. John S. Partridge, and cashier Charles M. Blackman. The bank officially opened for business on January 2, 1864. The First National Bank moved to the corner of Main Street and First Street in 1892. This photograph shows the bank's lobby after renovations in 1893. (First Citizens State Bank.)

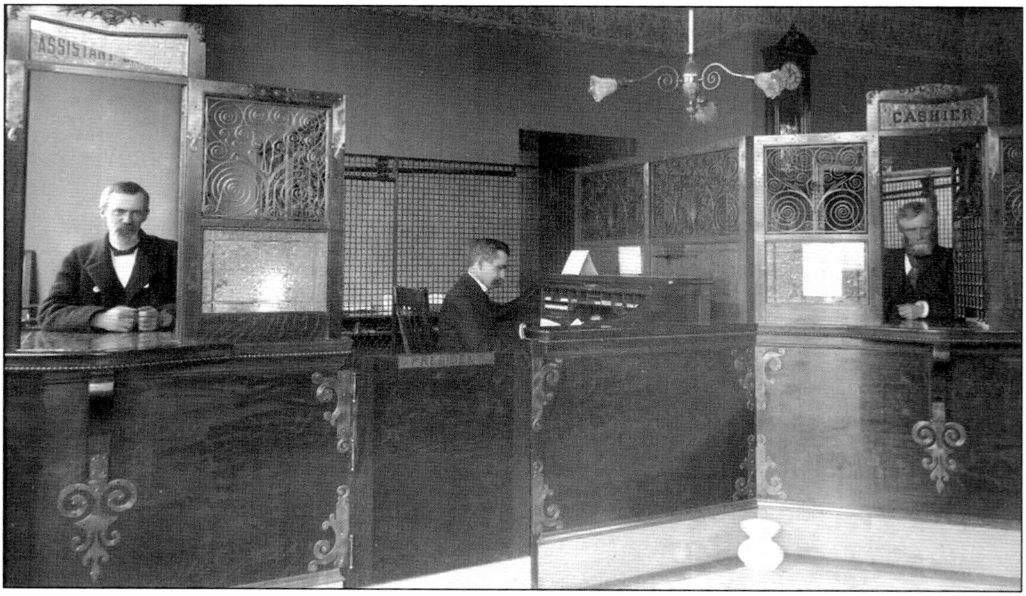

This 1894 photograph depicts Charles M. Blackman (center) and cashiers Isaac U. Wheeler and Charles Wheeler in the Citizens National Bank of Whitewater. The original bank was located at 173 West Main Street. Blackman was bank president, and the Wheelers were longtime employees. The cashiers kept large leather books or ledgers to record each account transaction. (First Citizens State Bank.)

Charles M. Blackman was president of the First National Bank of Whitewater for 40 years. He succeeded Sanger Marsh after the original president's untimely death in a runaway horse and carriage accident. After Blackman's death in 1912, his son Thane M. Blackman became bank president. The bank was located at Center Street and First Street at this time. (First Citizens State Bank.)

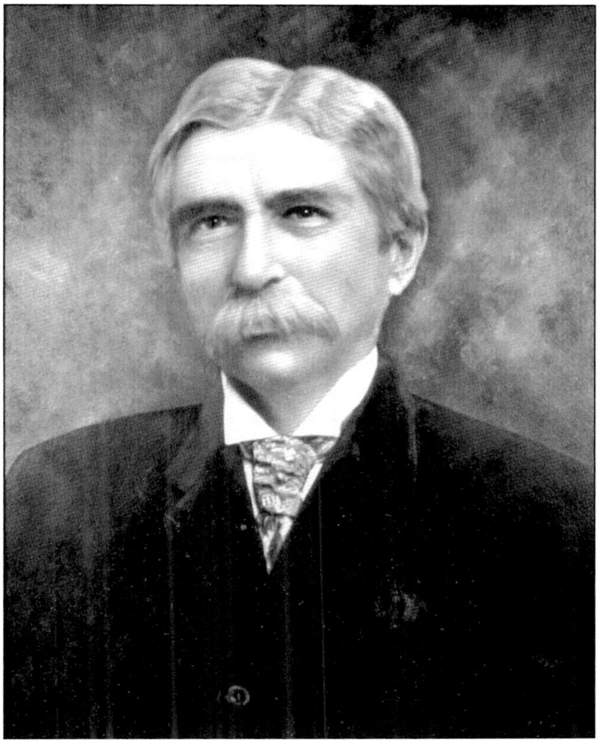

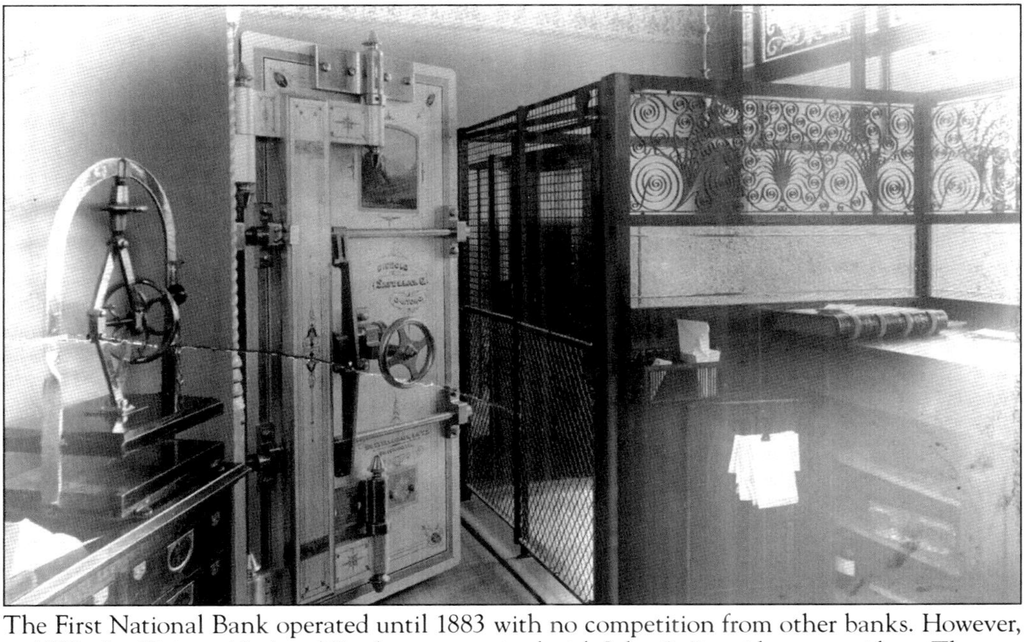

The First National Bank operated until 1883 with no competition from other banks. However, in 1883, the Citizens National Bank was organized with John S. Partridge as president. The two rival banks later merged in 1894, when Citizens National Bank president E. M. Johnson wanted to retire but was unable to sell the bank due to national financial troubles. This 1894 photograph shows the vault of the First National Bank. (First Citizens State Bank.)

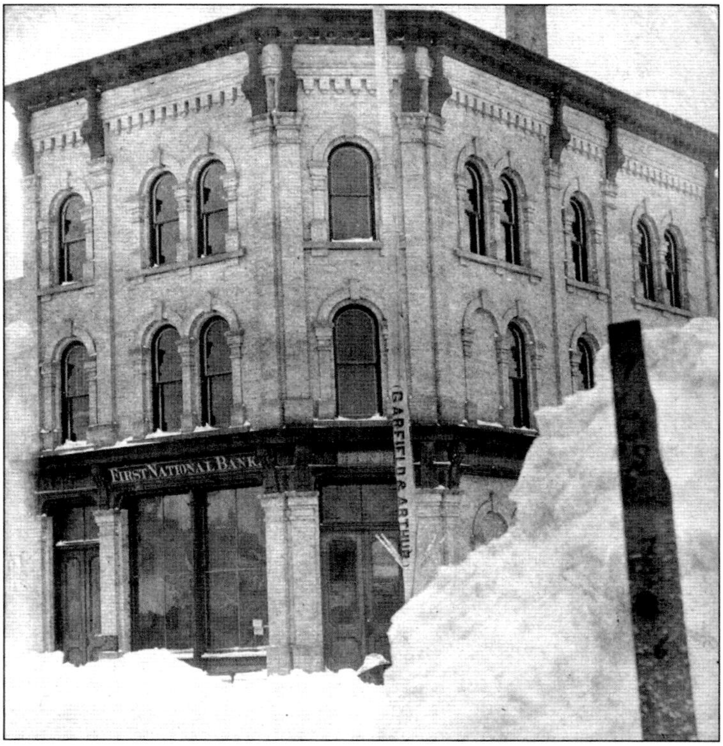

The First National Bank moved to this location on the corner of First Street and Center Street in 1874. The bank is shown here after the snowstorm of March 2–4, 1881, that made much of Whitewater impassible, buried under huge drifts of snow up to 20 feet high. The pole in front of the bank advertises the presidential ticket of "Garfield & Arthur." (WHS.)

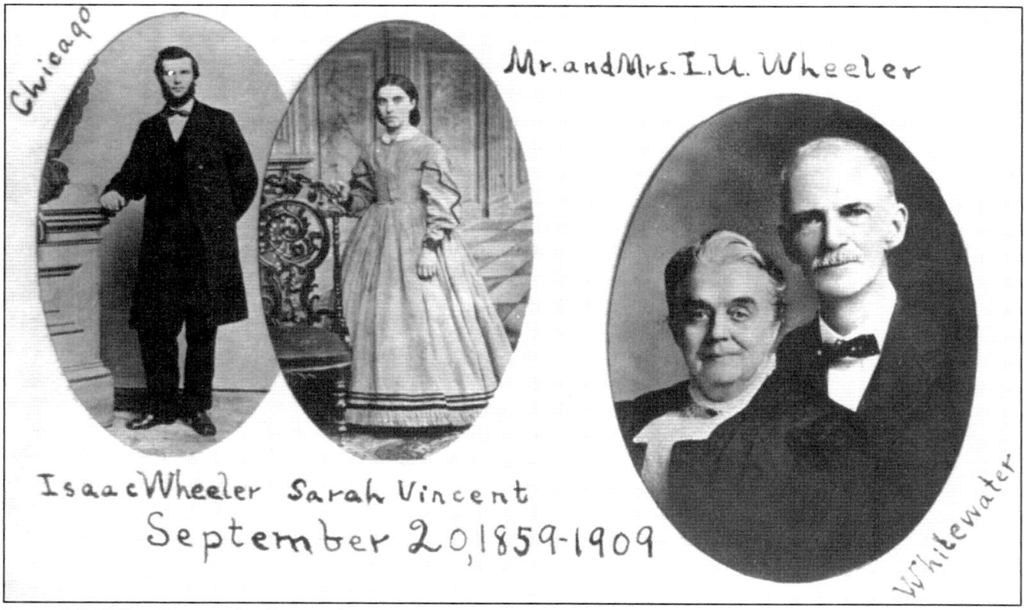

Isaac U. Wheeler worked in Whitewater's banking institutions for more than 50 years. He was first a clerk in an independent bank, starting in 1859. He then served as a bookkeeper for the First National Bank from 1868 to 1882, and then as assistant cashier until 1898. He then became a cashier at the Citizens State Bank. Pictured is a 50th anniversary card commemorating his marriage to Sarah Vincent. (First Citizens State Bank.)

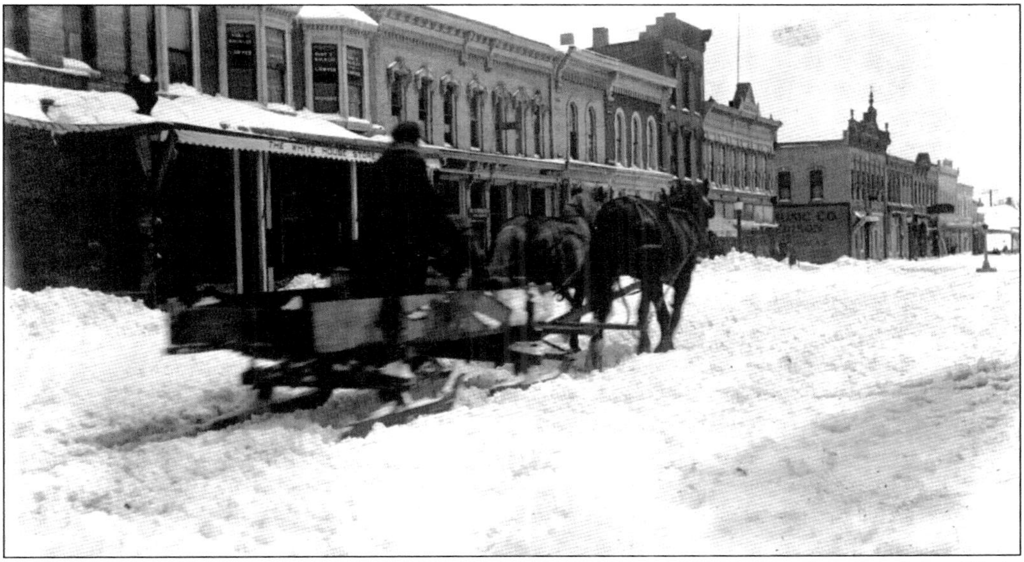

The Whitewater Cutter Works opened in 1876 under Joseph H. Page, Leonard C. Smith, and J. A. Chapman. The business manufactured cutters, sleighs, and buggy tops. It made 400–600 cutters and 10,000 buggy tops each year and employed about 20 people. Cutters and sleighs were essential in the winter months for transportation since travel was easier when unpaved roads were covered with packed snow and frozen. (Irvin L. Young Memorial Library; Fran Achen Collection.)

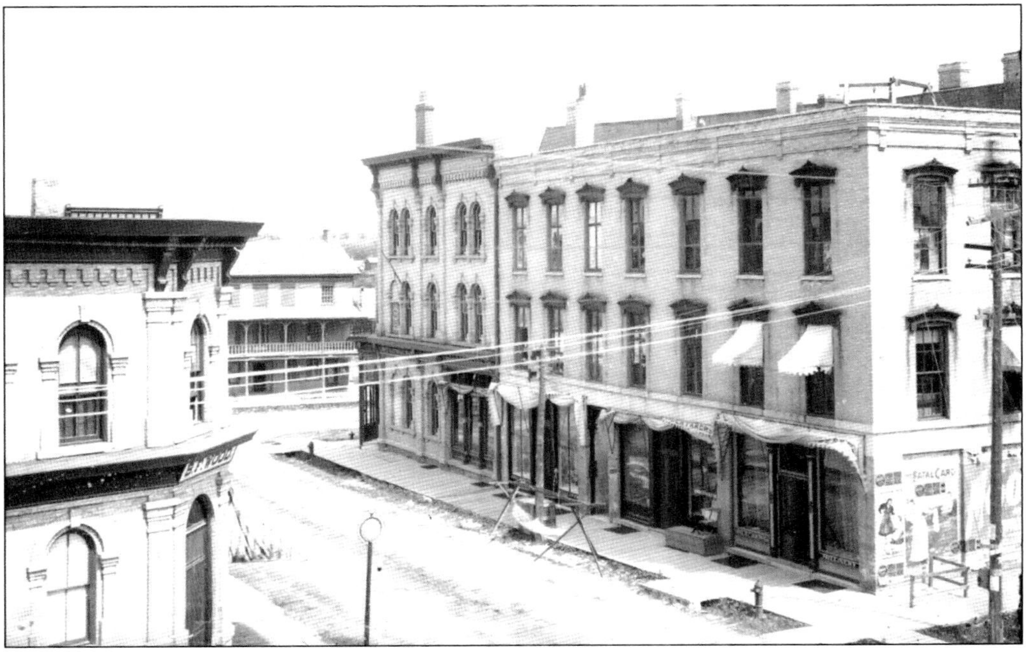

Benjamin F. Wood operated a jewelry shop, pictured at left in about 1916, on the corner of First Street and Center Street. He first opened for business in 1880 and offered silverware, clocks, watches, and optical goods. He manufactured his own high-quality violins and carried a line of small musical instruments in his shop. (Irvin L. Young Memorial Library; Fran Achen Collection.)

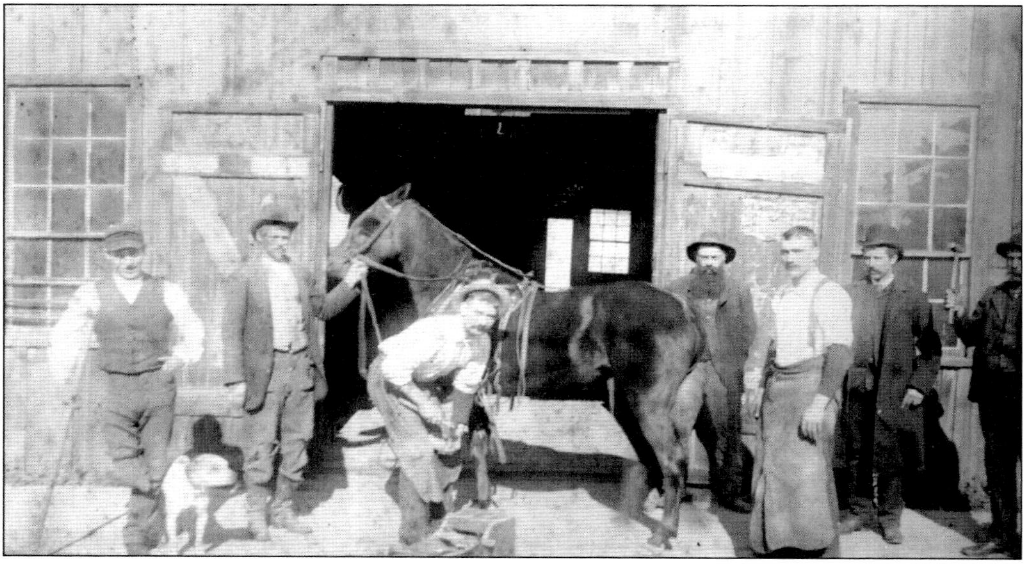

This photograph was taken outside the John Malone Blacksmith shop in 1900. The shop was located on the corner of Whitewater Street and Fourth Street. Timothy H. Dorr, a longtime Whitewater resident, is in the center of the photograph, shoeing the horse. John Malone is to the right with white shirt and suspenders. (Irvin L. Young Memorial Library; Fran Achen Collection.)

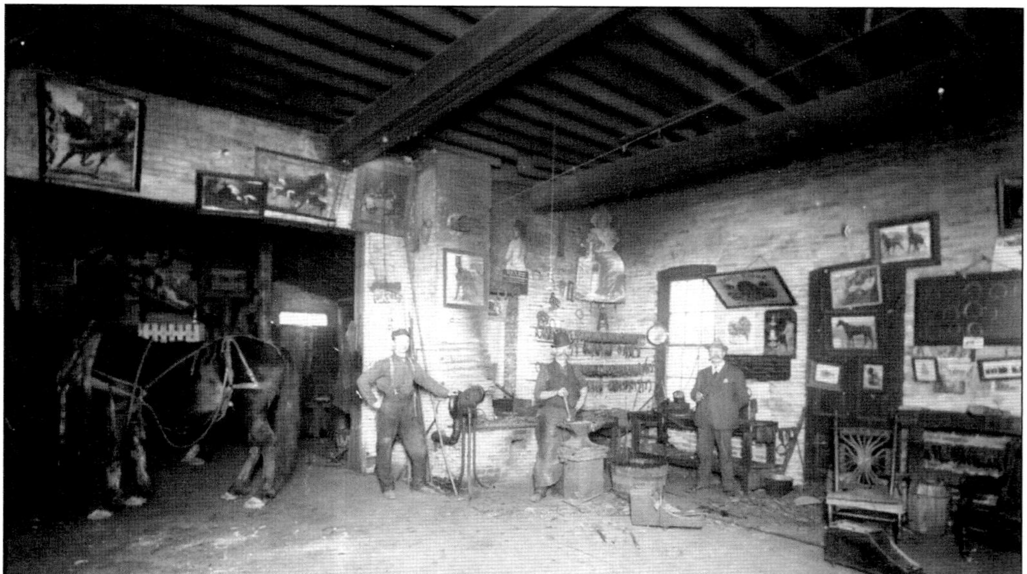

Timothy H. Dorr's blacksmith shop, shown here in 1907, first opened in 1880 at 111 Whitewater Street. Pictured are teamster Boyd Rasmusson, Timothy Dorr, and C. H. Leffingwell. During the 1880s, Whitewater had seven blacksmith shops that performed all manner of tasks, including iron repair, shoeing horses, building barrels and wagons, and creating iron fencing. (Irvin L. Young Memorial Library; Fran Achen Collection.)

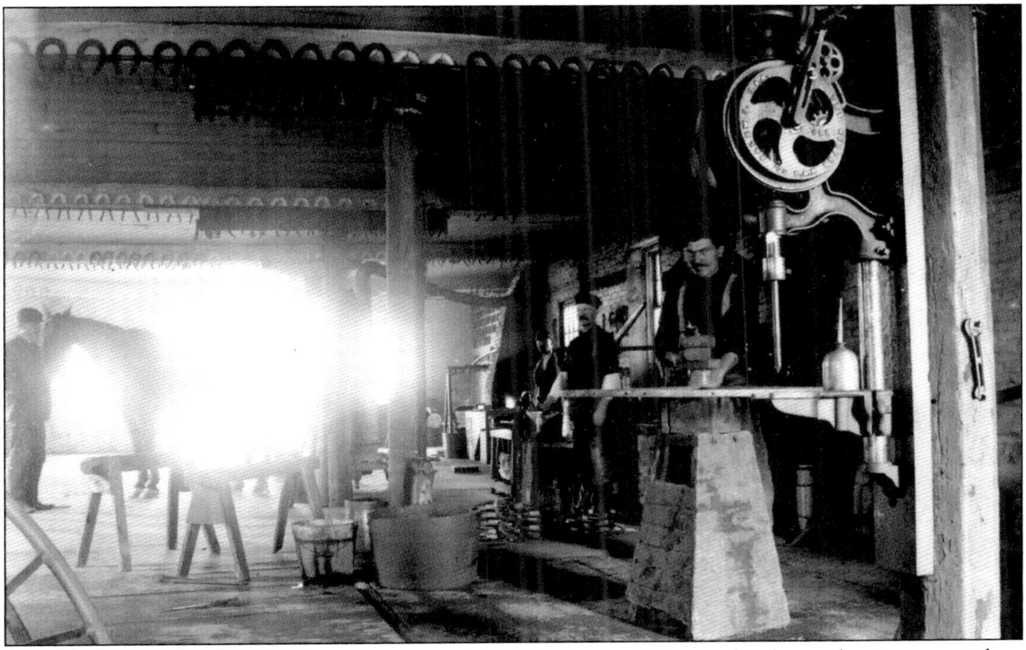

The Olmstead Blacksmith Shop was owned by W. Harvey Olmstead, a horseshoeing specialist. Blacksmithing was a common trade in the 19th century. However, the increased mechanization of American factories meant that many goods formerly produced by a blacksmith could be mass-produced, reducing the importance of the blacksmith in his community. (Irvin L. Young Memorial Library; Fran Achen Collection.)

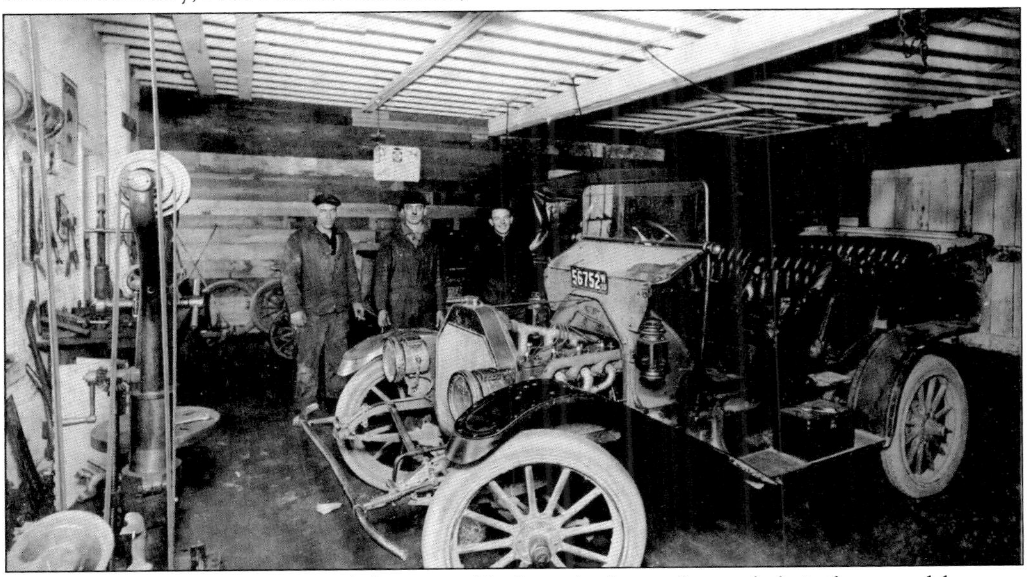

With the advent of the automobile, many blacksmith shops changed their focus and became garages to repair automobiles in addition to farm wagons and other metal products. Early garages washed, stored, and repaired cars. The more extensive garages even offered automobile rental and chauffeur services. Edwin M. Drewry, Clarence Strutz, and John E. Dorr are pictured here in 1917 in the Dorr and Drewry Garage. (Irvin L. Young Memorial Library; Fran Achen Collection.)

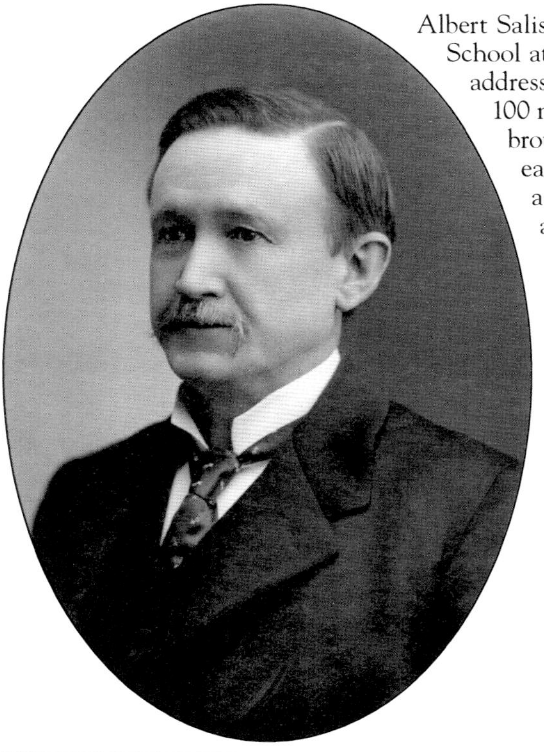

Albert Salisbury was the president of the State Normal School at Whitewater from 1885 to 1912. In an 1887 address, he claimed that the school had brought 100 new families to Whitewater and that it brought a net profit of nearly $50,000 to the city each year of operation. The school provided a consistent source of employment and attracted the most intelligent young people from surrounding towns. (WHS.)

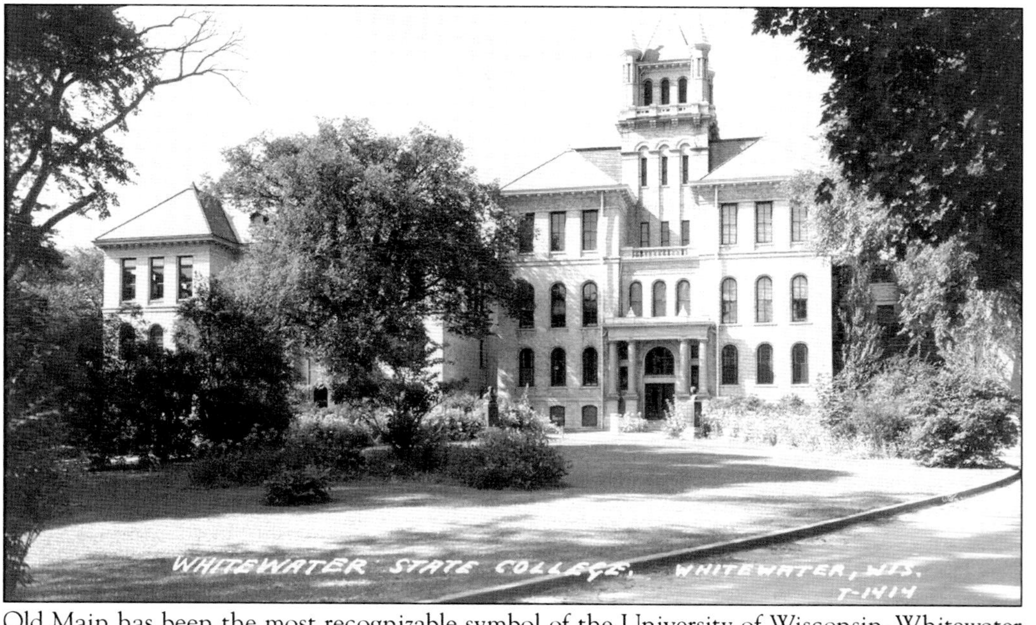

Old Main has been the most recognizable symbol of the University of Wisconsin–Whitewater since its construction in 1868. The Whitewater Normal School was founded as a teacher's college in 1868, with Newton M. Littlejohn and Judge Samuel Austin White in charge of the construction committee. Whitewater competed with many other sites for the privilege of hosting the second State Normal School. (Russ Fero.)

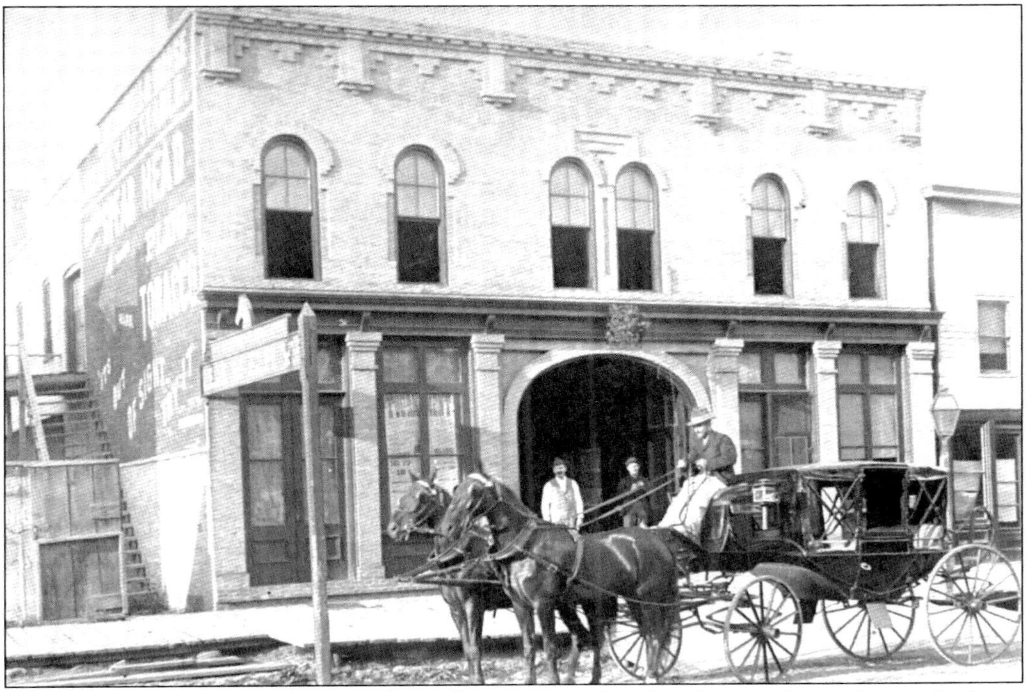

Leffingwell's Livery, pictured, was located on the north side of Main Street. It was built on the site of the former Montour House, operated by Septer Wintermute. After the Montour House burned down in 1865, Wintermute built this livery barn, which was run by his grandson Harry C. "Chap" Leffingwell. (Irvin L. Young Memorial Library, Fran Achen Collection.)

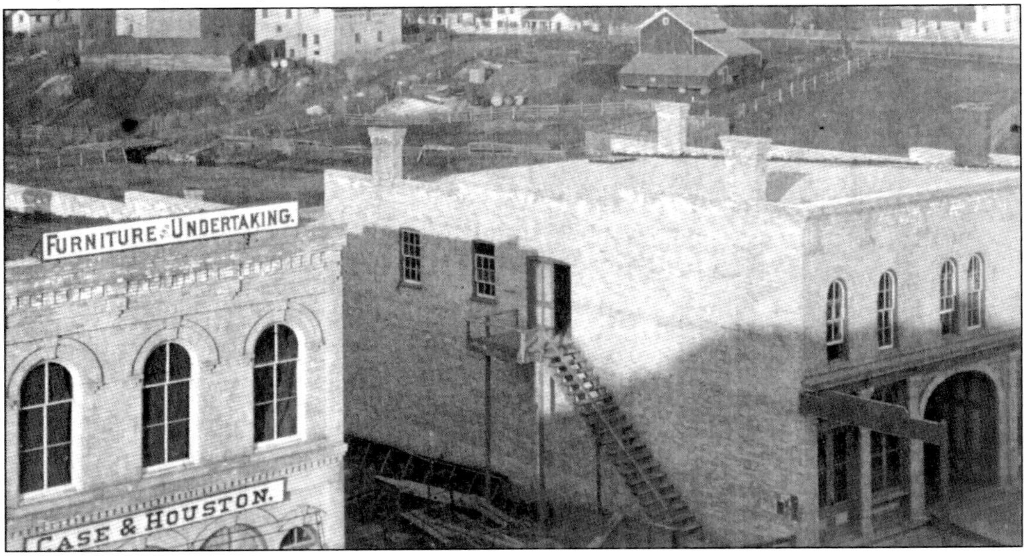

James M. Case and John Houston established Case and Houston in 1875. They sold chairs, tables, bedsteads, chamber suites, and parlor suites. They also stocked coffins and caskets and served as undertakers. Case and Houston was located on Main Street, next to Leffingwell's Livery. The white building in the distance is the Whitewater Brewery. (Irvin L. Young Memorial Library; Fran Achen Collection.)

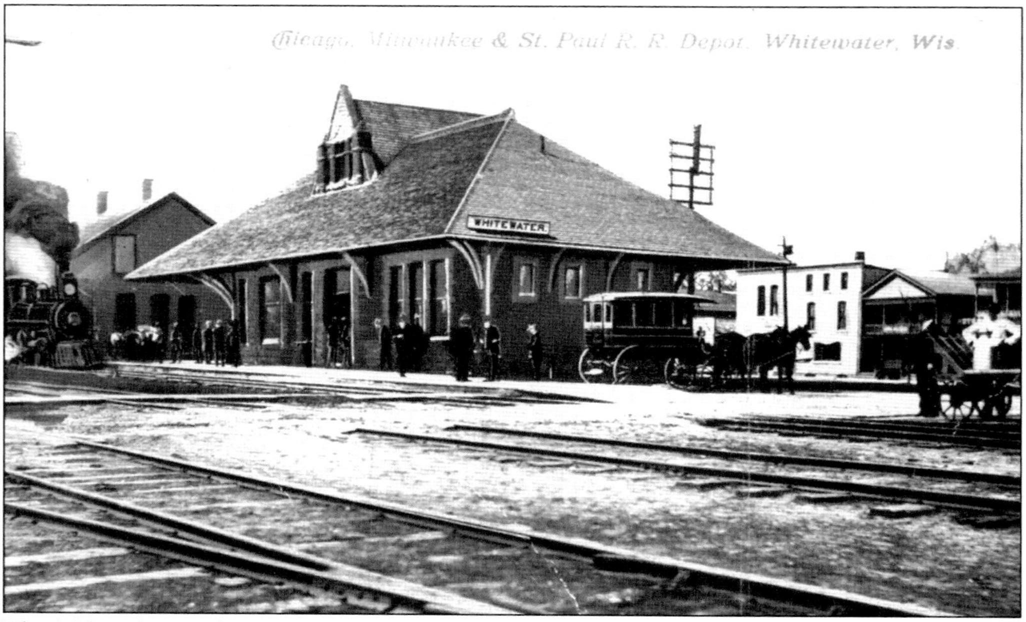

The Milwaukee and Mississippi Railroad was the biggest contributor to Whitewater's growth. After the railroad was completed in 1852, Whitewater's population doubled in three years. In 1857, Whitewater was the busiest of the 14 stations on the railroad line. A third of the wheat and barley shipped over the course of the railroad came from Whitewater, as did a quarter of the wool. (Irvin L. Young Memorial Library.)

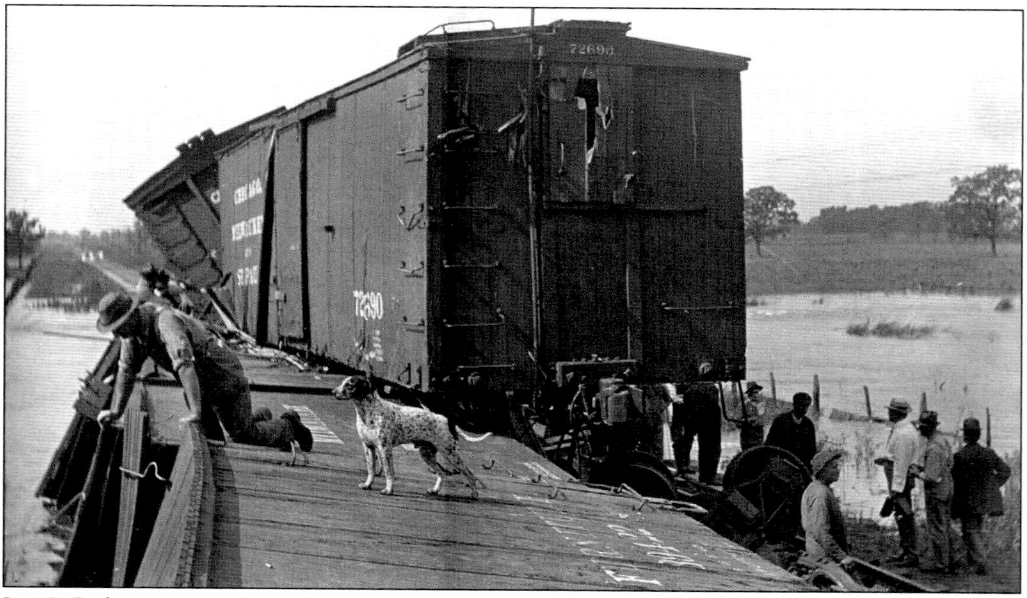

In 1915, this engine of the Chicago, Milwaukee and St. Paul Railroad (formerly the Milwaukee and Mississippi) ran off its track. The wreck happened between Whitewater and Palmyra. Rails manufactured and laid in the early 1900s were not designed to withstand larger, heavier rail cars, leading to many accidents. It appears that this particular wreck was a popular spectacle. (Irvin L. Young Memorial Library; Fran Achen Collection.)

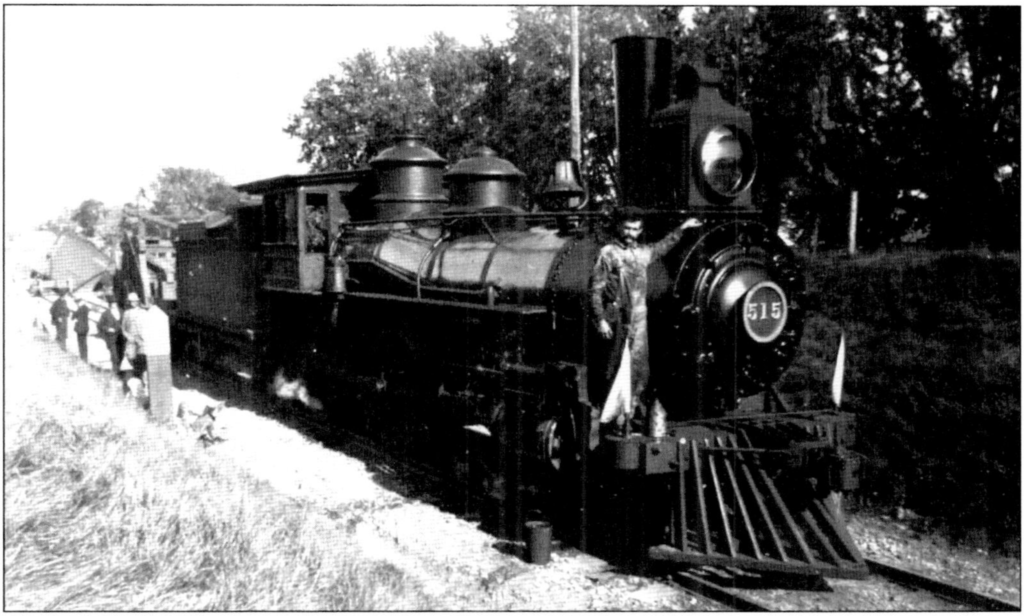

This photograph shows the 515 engine of the Chicago, Milwaukee and St. Paul. The railroad ran 196 miles from Milwaukee to the Mississippi River. Its construction was partially funded by the Walworth Country farmers who mortgaged their farms and contributed what they could. By 1898, Whitewater's Station Master, Charles C. Danforth, welcomed 8 to 10 trains a day and sold more than 6,000 passenger tickets annually. (Irvin L. Young Memorial Library; Fran Achen Collection.)

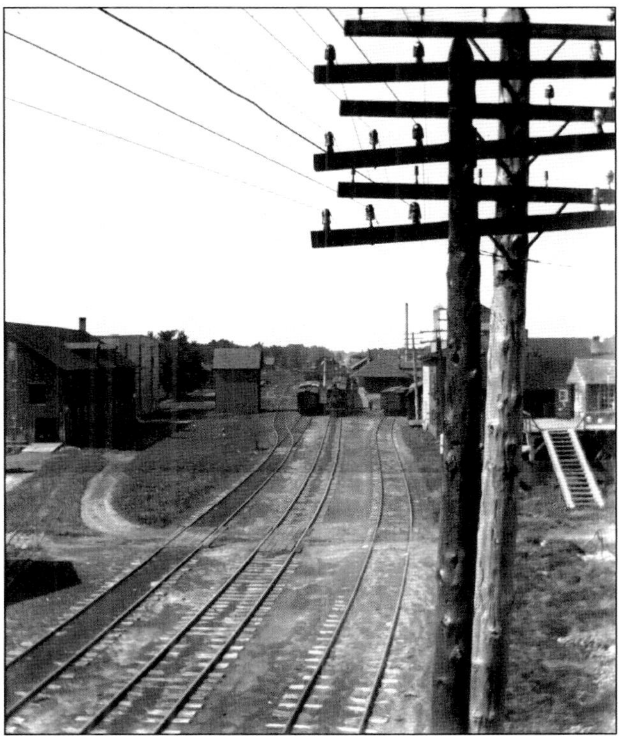

The railroad yard was located south of the Chicago, Milwaukee and St. Paul depot, along James Street. Here the rails diverged for easier loading of goods from the warehouses and grain elevators that lined the tracks. Many of Whitewater's notable businessmen also owned warehouses. As early as 1854, Whitewater had seven grain warehouses, and by 1861 there were 14 warehouses that lined the south side of Whitewater Street. (Irvin L. Young Memorial Library; Fran Achen Collection.)

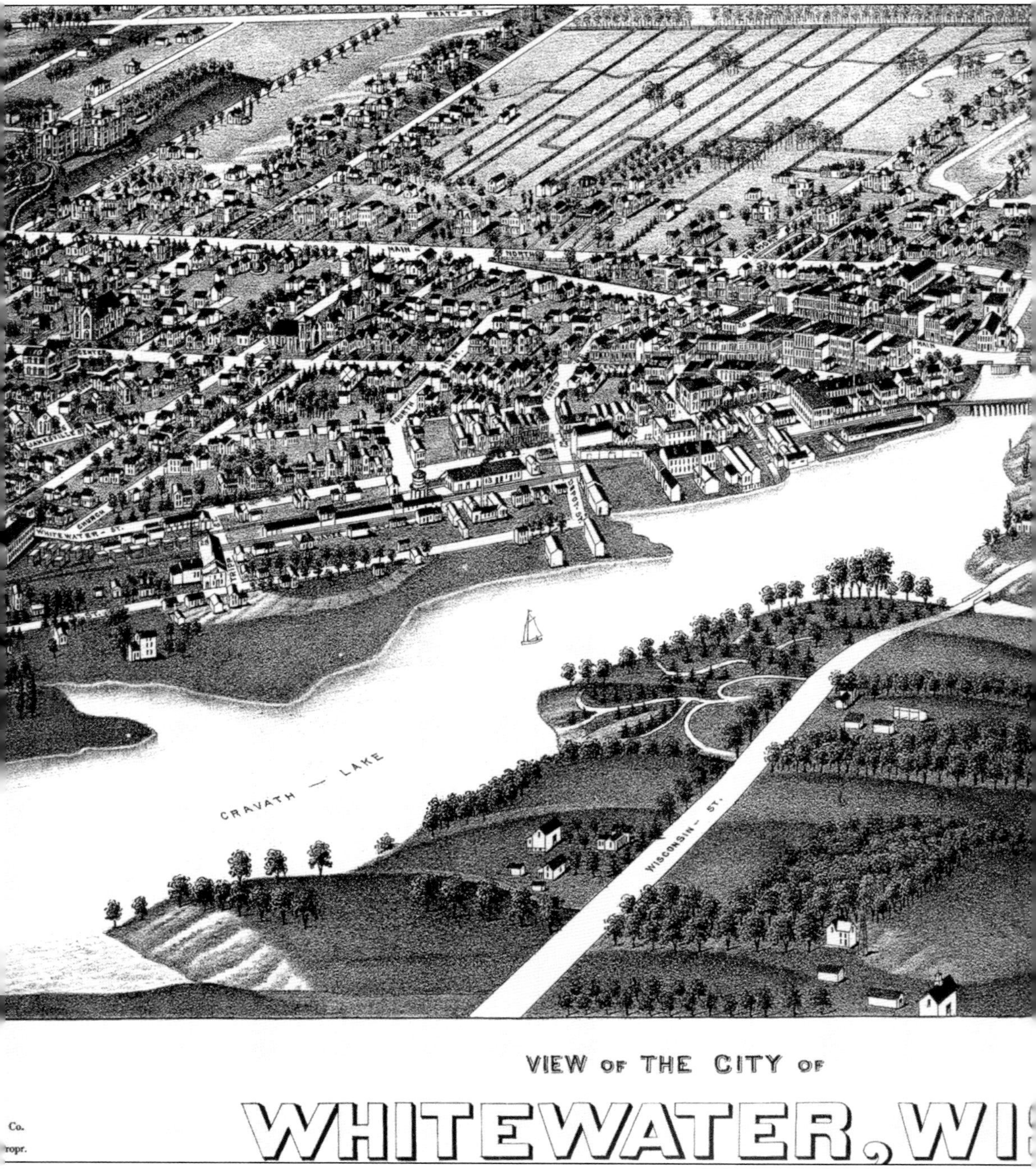

This map shows Whitewater in 1885. As early as 1869, some Whitewater residents argued that the town had enough money to "establish a half dozen good sized manufacturing establishments." One proposal called for a woolen factory, monthly fairs, tanneries, and a boot and shoe factory, but no such large establishments developed. An 1869 editorial in the *Whitewater Register* now appears particularly prophetic: "We know there are those who will say that Whitewater has steadily gone forward; but we would remind them that such progress has been due, mainly, to

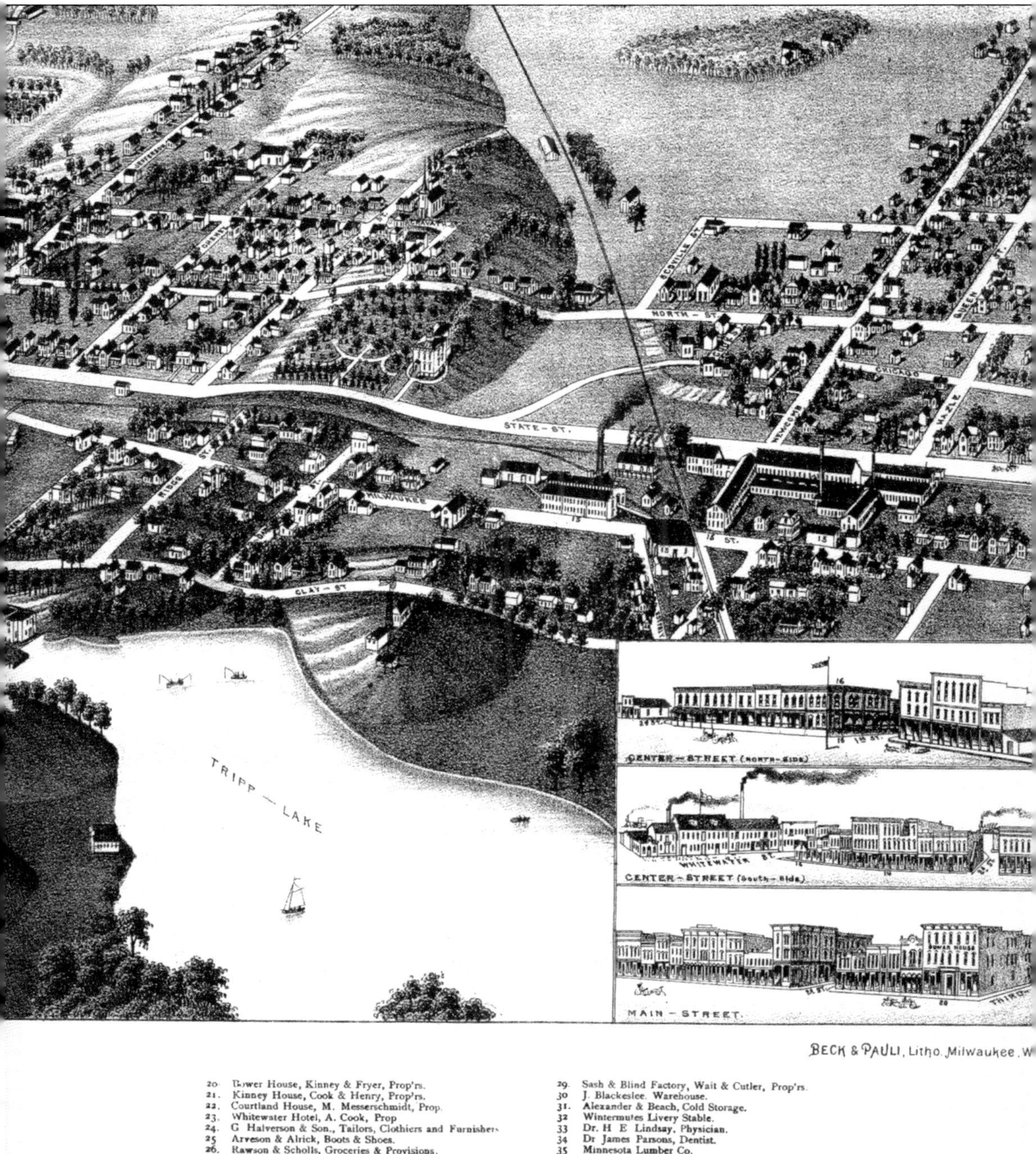

the manufacturing interests we already have established among us. Take away the factories of Geo. Esterly, Winchester, Partridge & Co . . . and what would happen to us? One-half of the merchants would leave at once, the printing office would seek other fields, and the grass would grow in Main Street." Little heed was paid to these warnings. Whitewater coasted through the 1870s and 1880s with few real changes, unaware of what was soon to come. (WHS.)

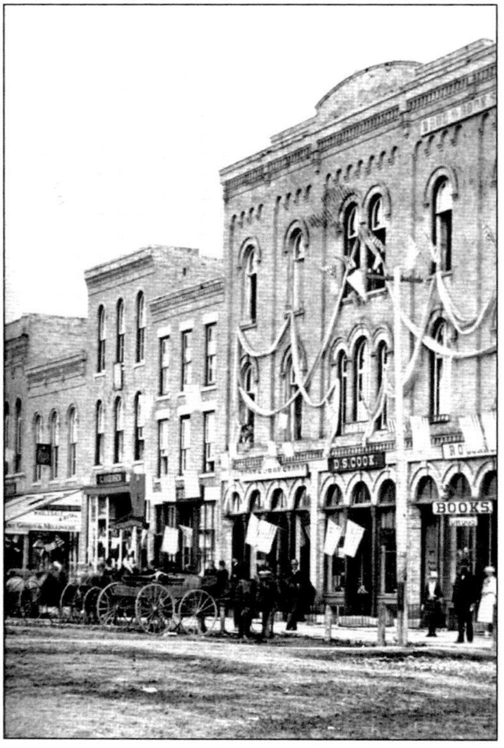

This photograph shows the Main Street dry goods shop of D. S. Cook. Daniel Seymour Cook came to Whitewater in the fall of 1849. He began working for Peck, Keep, and Company in the mercantile business. Cook went into business with Peter H. Brady and Jacob J. Starin in 1851. By 1863, Cook was the sole owner of the business, which sold dry goods, groceries, hardware, boots, and shoes. (WHS.)

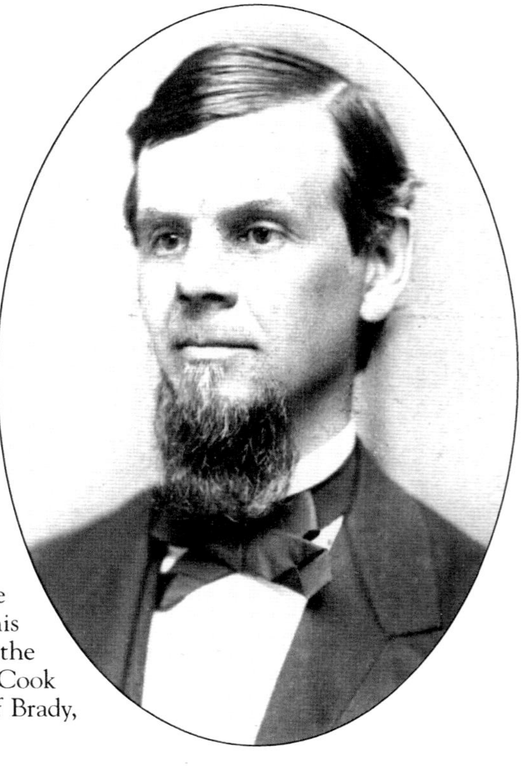

Daniel Seymour Cook was a longtime resident of Whitewater who was actively involved as clerk of the Congregational society for more than 50 years, starting in 1853. He and his wife Rachel were also generous supporters of the Whitewater City Library. His brother Lewis Cook joined him in business after the dissolution of Brady, Starin and Cook in 1857. (WHS.)

The Cook home, pictured, was built in 1856. It was located on the north side of Main Street on property that now belongs to the University of Wisconsin–Whitewater. Notice the unique design of the tower on the Italianate-style home, which was torn down to make way for the Andersen Library. Daniel Cook is visible on the front porch. (WHS.)

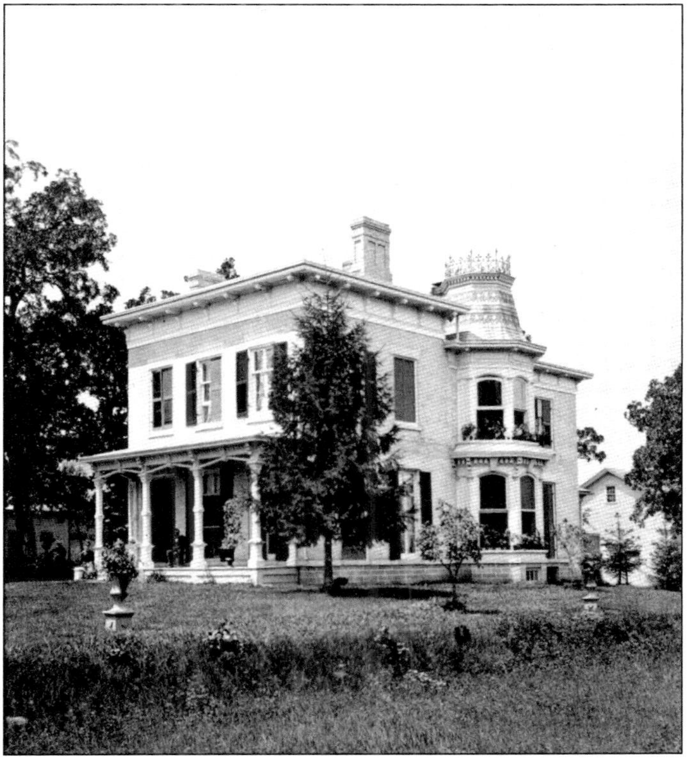

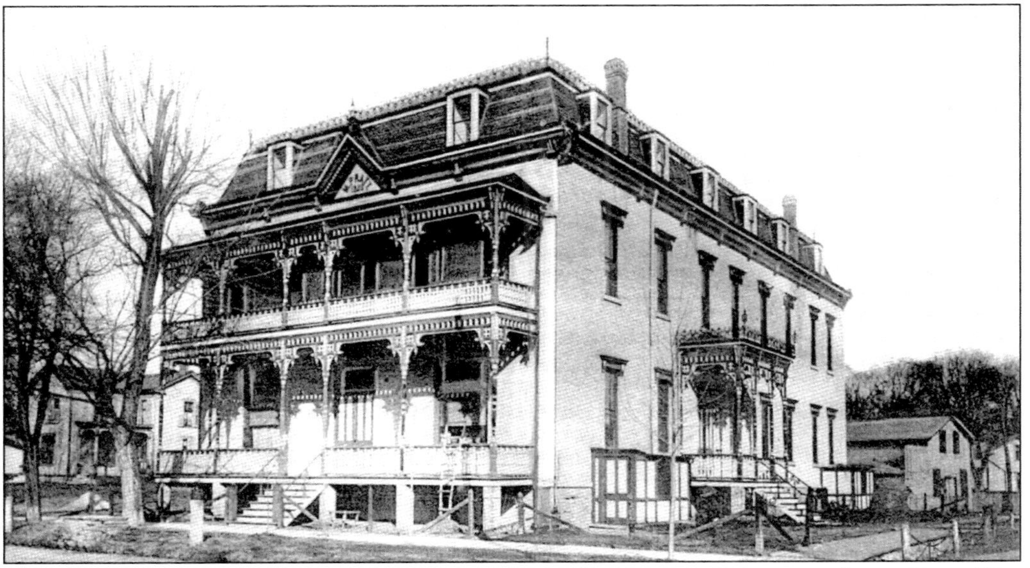

This photograph shows the Morris Pratt Institute, the nation's first spiritualist college. The much-misunderstood institute did not deal in witchcraft. Rather, its curriculum included history, math, composition, and other traditional disciplines, as well as classes on psychic research and spiritualism in the Bible, with "the desired aim to make all students thinkers." The institute was also the site of plays, town meetings, lectures, and raucous public séances. (Irvin L. Young Memorial Library.)

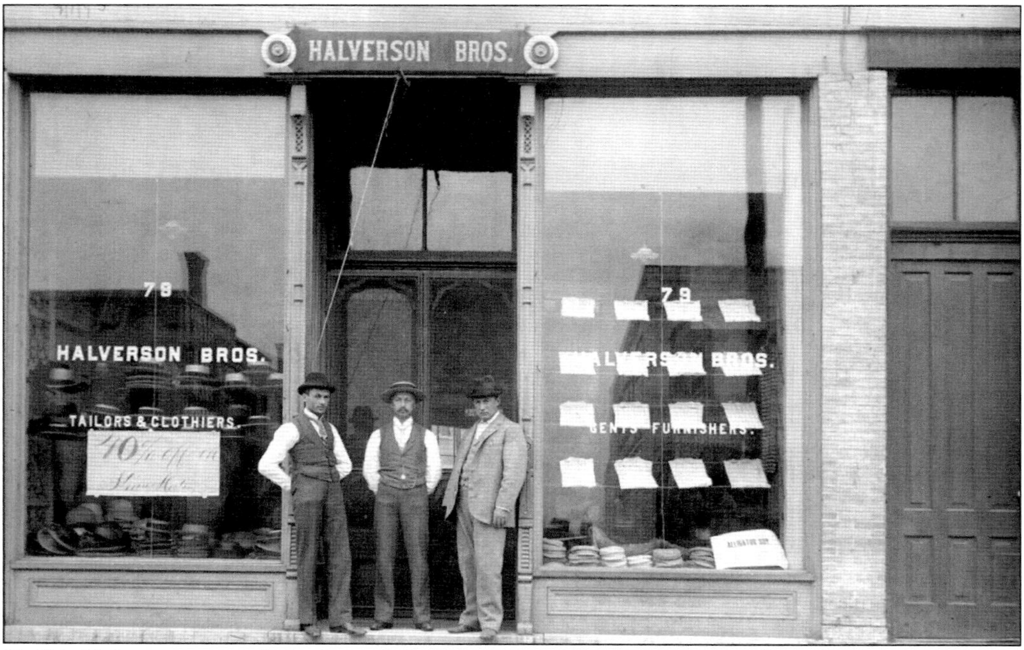

Gullik Halverson started the firm of G. Halverson and Son, Tailors and Clothiers in 1881, with his son Miner. Halbert, Gullik Halverson's youngest son, became a clerk in the store. Eventually the two brothers formed Halverson Brothers Clothiers and Furnishers, pictured here. Previously Gullik had formed a partnership with Gilbert Andersen in 1863 that lasted until 1873. This 1893 photograph depicts Halbert Halverson, left, and Miner in the middle. (WHS.)

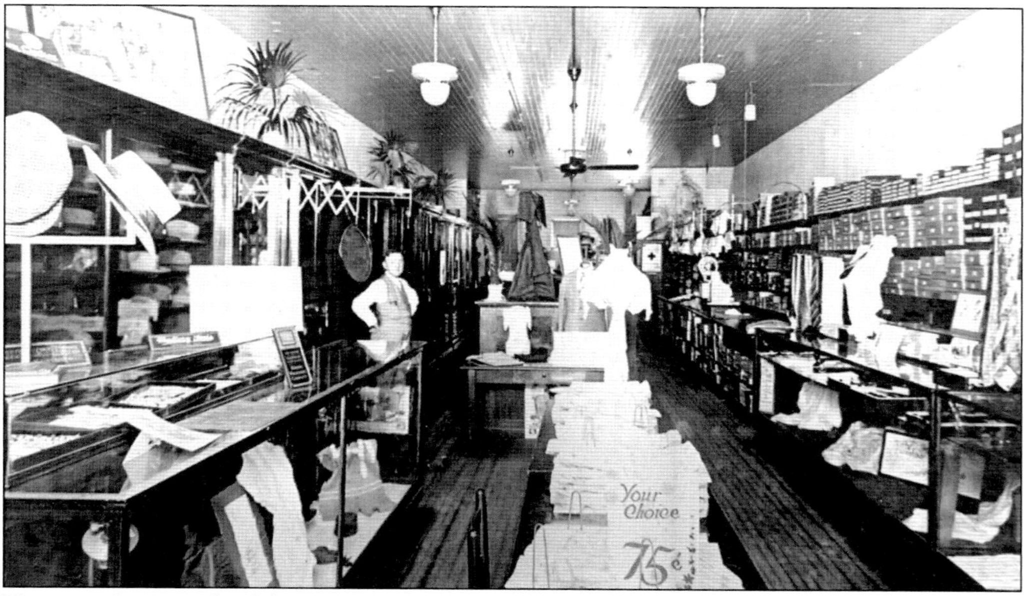

This is a photograph of the interior of Halverson Brothers Clothiers and Furnishers. The store had many modern amenities, such as electric lights and an electric fan extending from the ceiling. On the table in the middle is a display of shirts for a mere 75¢ apiece. Halverson Brothers also made its own hats. (Irvin L. Young Memorial Library; Fran Achen Collection.)

The Halverson men were bankers as well. Miner G. Halverson, pictured here, his grandfather Newton Littlejohn, and his son Donald Halverson were all directors of the First Citizens National Bank at one time. In addition to the clothing store, Miner held interest in other business ventures, such as the Union Produce Company. (WHS.)

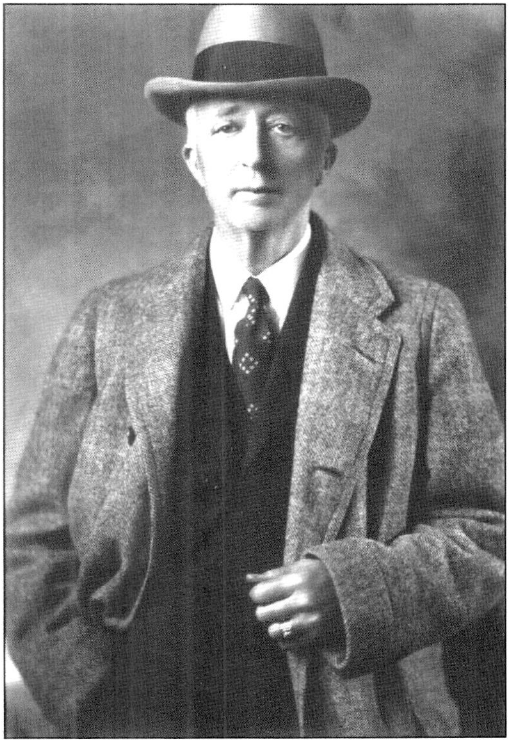

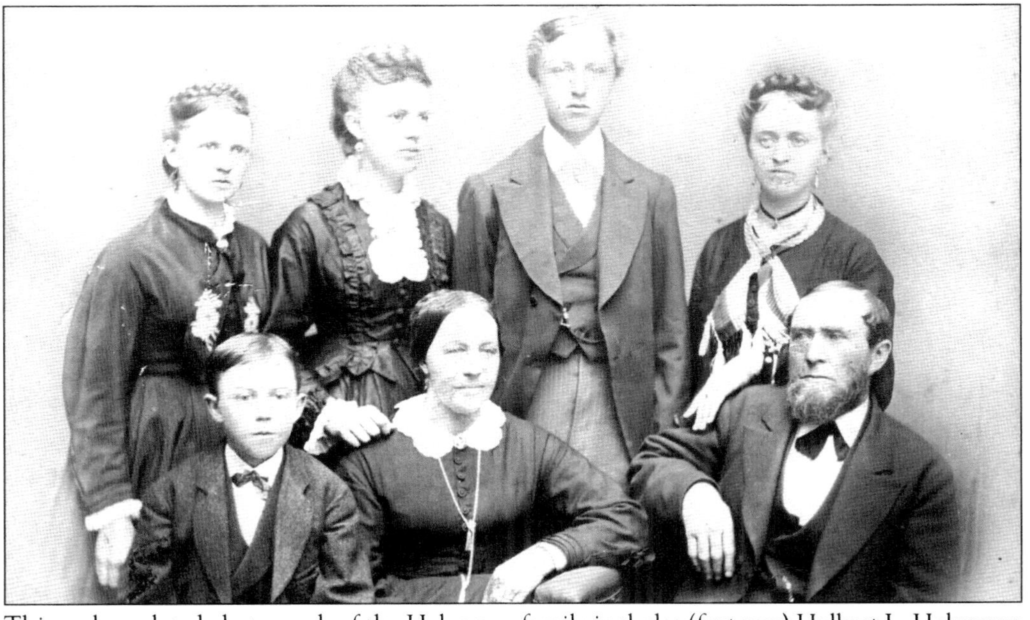

This early undated photograph of the Halverson family includes (first row) Halbert L. Halverson (husband of Maude Bovee), Dorothea Halverson (wife of Gullik), and Gullik Halverson; (second row) Mary Mason (wife of Ingbert Mason), Julia Rosman (wife of Edmund Rosman), Miner G. Halverson (husband of Mary Littlejohn), and Catherine Simonson (wife of Simon Simonson). (WHS.)

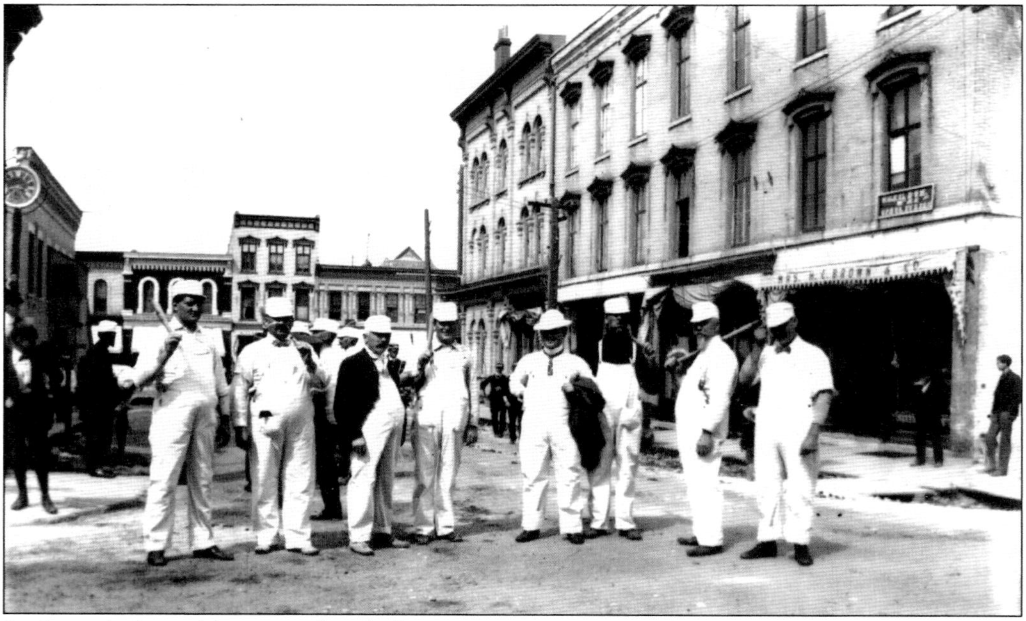

In the early days of American baseball, many communities established baseball contests between "the Fats" and "the Leans," in which a "hefty" team of local businessmen would compete against a "slim" team. These games were often held as charitable events, as enthusiastic crowds paid to watch the spectacle unfold. These men, participants in Whitewater's version of the game, are pictured at the end of First Street. (Irvin L. Young Memorial Library; Fran Achen Collection.)

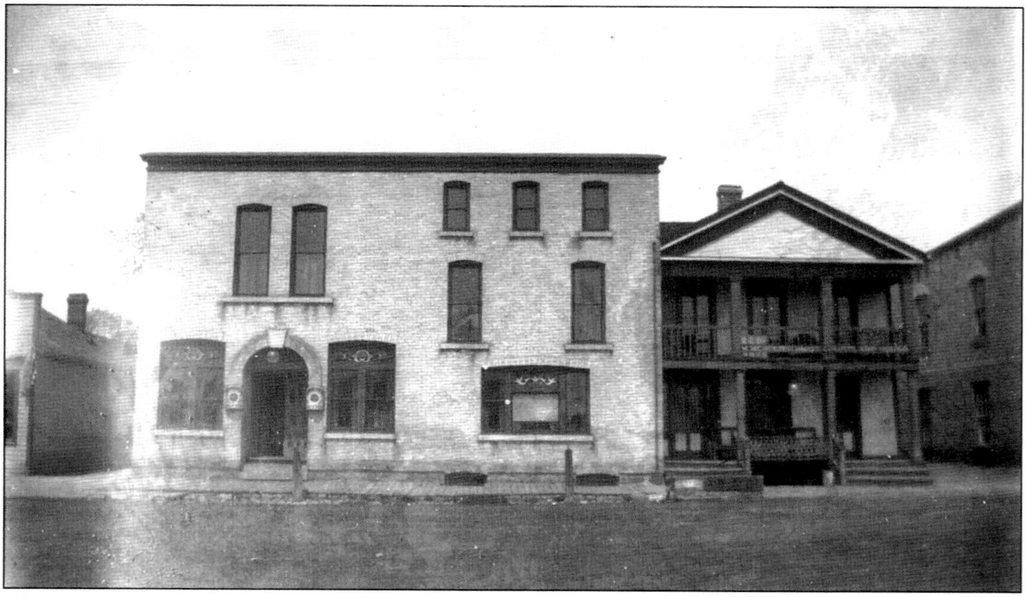

The Cortland House hotel was located on Whitewater Street. Constructed in 1854 as the Badger State Hotel, the building was improved in 1863 with a large brick addition and renamed the Cortland House. The Cortland House offered 28 rooms and remained until the 1960s, when the entire block was demolished to make room for Whitewater's new police and fire stations. (WHS.)

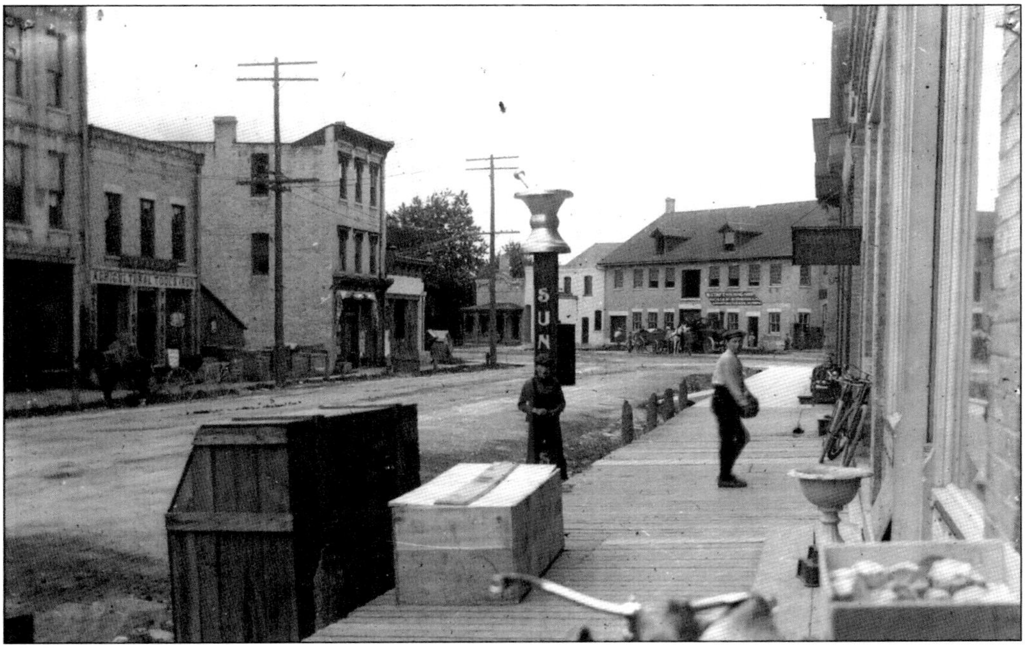

This view of Center Street, taken around 1900, shows the sidewalk outside the Sun Drug Store. The former Winchester and Partridge building lies in the background, occupied by Warren J. Taft's Machine Shop. The small sloped building at the left under the power lines was a public restroom. To the right is the office of Dr. Abram M. Leland. To the left is Frederick Finch's hardware store. (Irvin L. Young Memorial Library; Fran Achen Collection.)

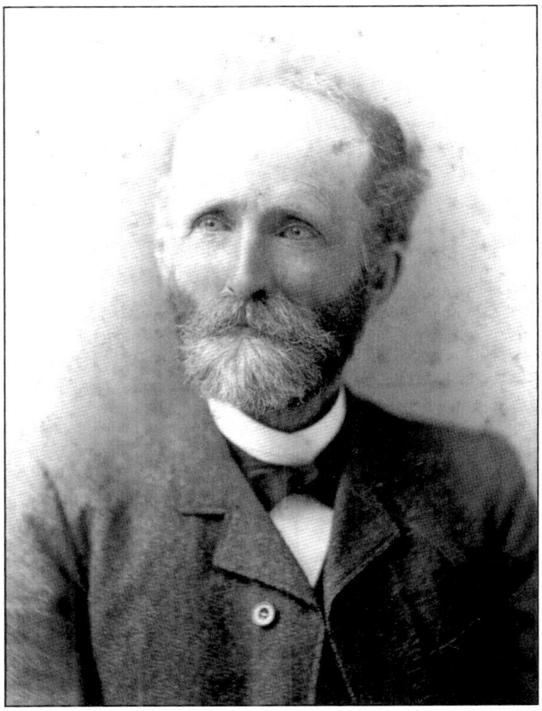

Dr. Abram M. Leland, pictured, was a homeopathic doctor. His father Aaron Granger Leland was also a physician, as well as secretary and local treasurer of the Milwaukee and Chicago Railroad Company before he finished his medical studies and moved to Whitewater in 1865. By 1882, there were 10 doctors listed in Whitewater, which had a total population of 4,523. (WHS.)

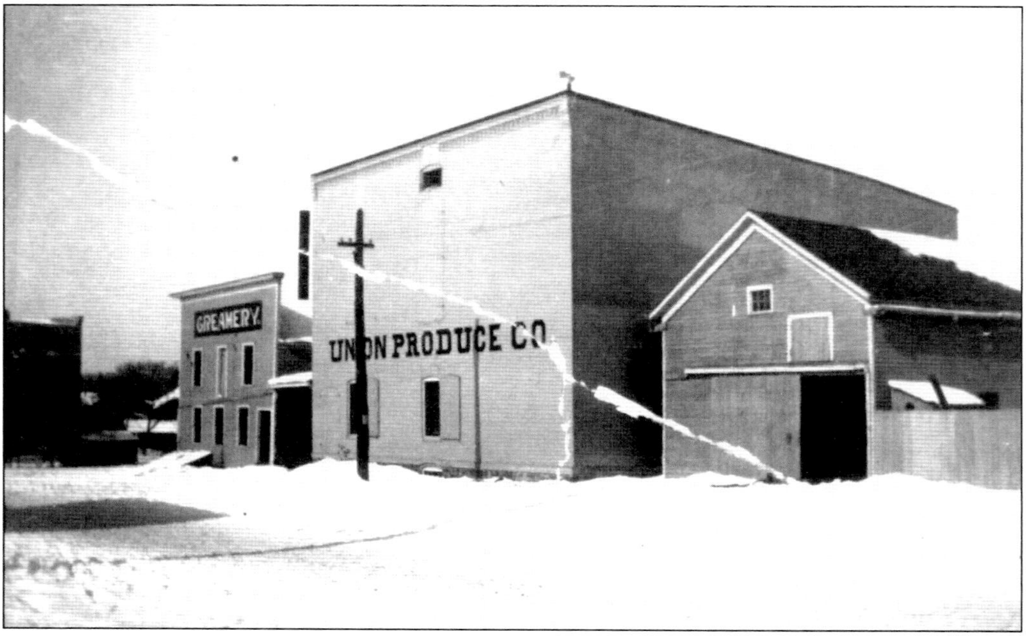

The Union Produce Company was organized in 1890, when William McCutchen built these facilities along Main Street. Avelyn I. Dexter was president of the company before former mayor Zadoc P. Beach took over. Beach, a graduate of the Whitewater Normal School, was considered to be "an expert in cold storage and creamery lines." (WHS.)

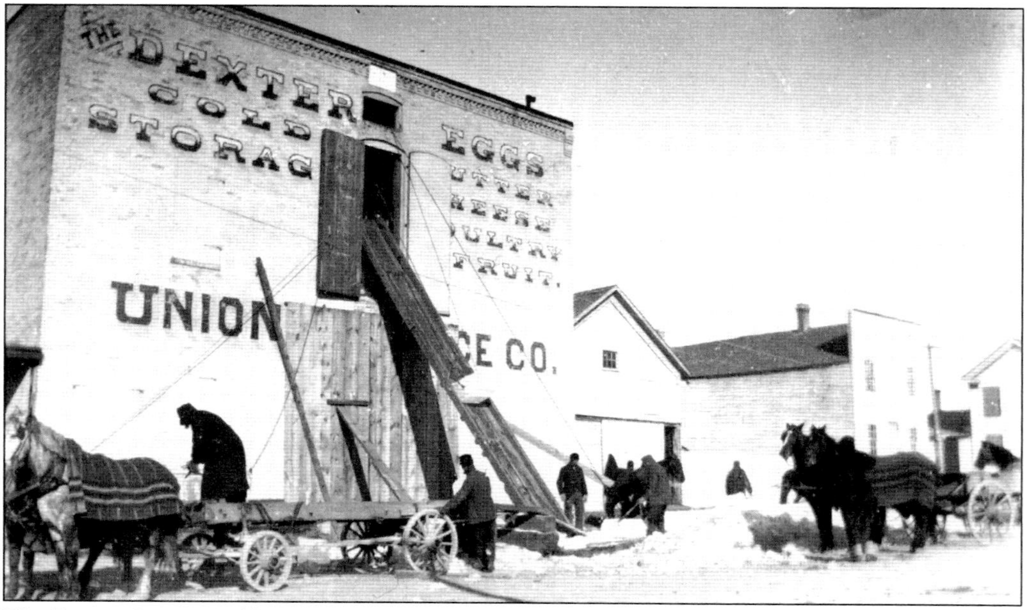

The Dexter Patent Cold Storage, invented by Avelyn I. Dexter, created a uniform temperature in each storage chamber year-round. Any moisture was eliminated through condensation. Dexter was president of the Union Produce Company, as well as a partner in the Whitewater Roller Mills. He patented his cold storage building system in 1894, a system that was used across the country. (WHS.)

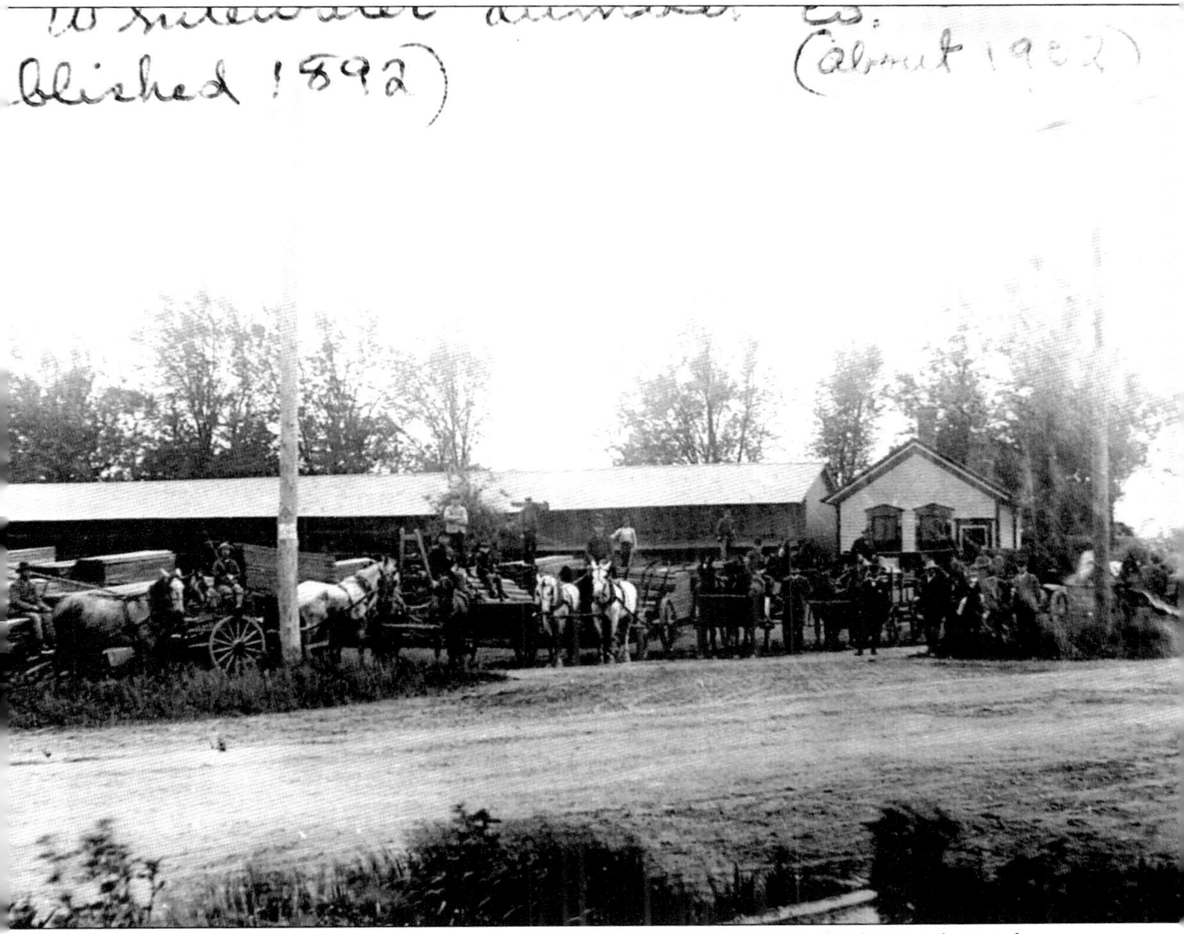

The Minnesota Lumber Company was headquartered in Polo, Illinois. It had more than a dozen branch shops across Wisconsin and purchased a large volume of lumber from local forests. The company came to Whitewater in 1883 and carried a full line of lumber and building materials, including laths, shingles, sashes, doors, blinds, and molding. It also carried a large line of coal of all kinds and grades. Jerome Baker managed the Whitewater outlet, which was taken over in 1892 by the Rockford Lumber and Fuel Company. The company was located on Wisconsin Street. (Irvin L. Young Memorial Library; Fran Achen Collection.)

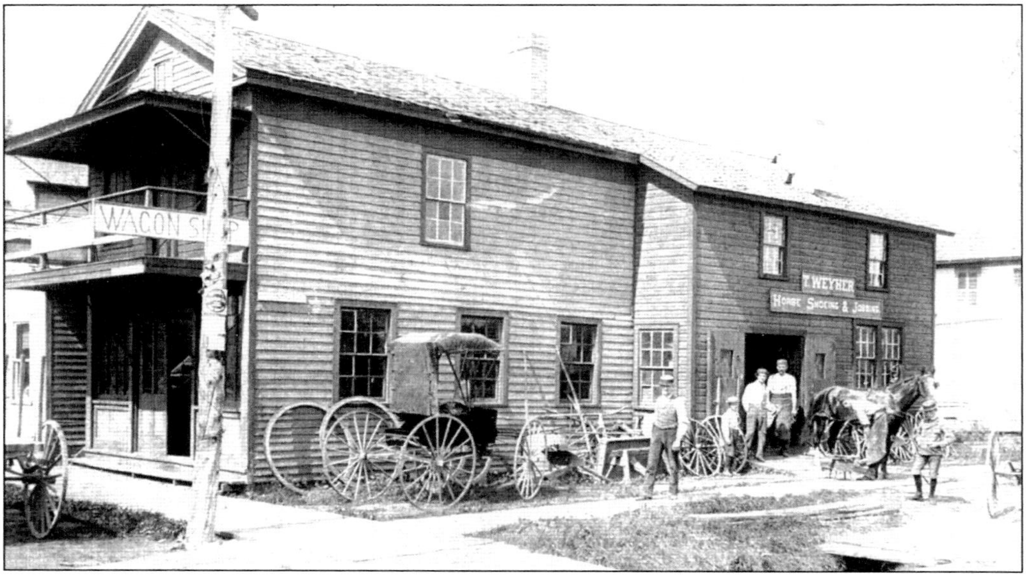

In 1890, blacksmith Theodore Weyher and wagonmaker Albert H. Marskie started building a light transportation wagon. They did "all kind of blacksmithing, horse-shoeing, buggy and carriage repairing in the most approved manner." Their building was located on the corner of Fourth Street and Whitewater Street. Marskie retired in 1895, and Weyher and his son Edward then started the Weyher Manufacturing Company, early Whitewater's last significant factory. (Henry Meisner.)

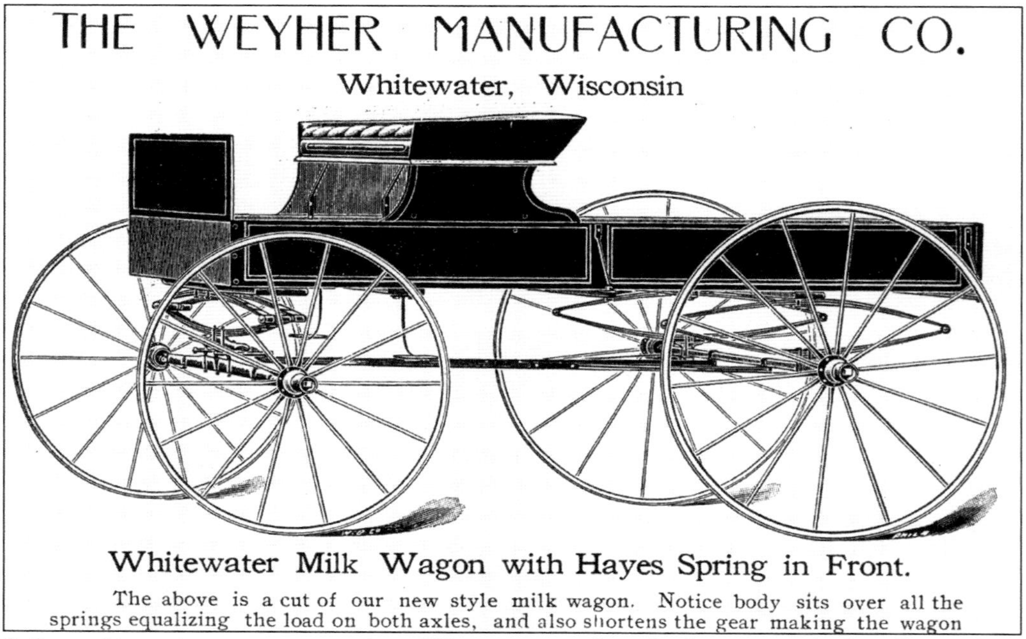

Before refrigeration was commonplace, Whitewater residents had to acquire fresh milk on a daily basis. Whitewater's numerous dairy farms competed for customers by offering home delivery in wagons like this one, produced by the Weyher factory. Milk was kept cool by storing the glass quart bottles in a tub of cold water. (Irvin L. Young Memorial Library.)

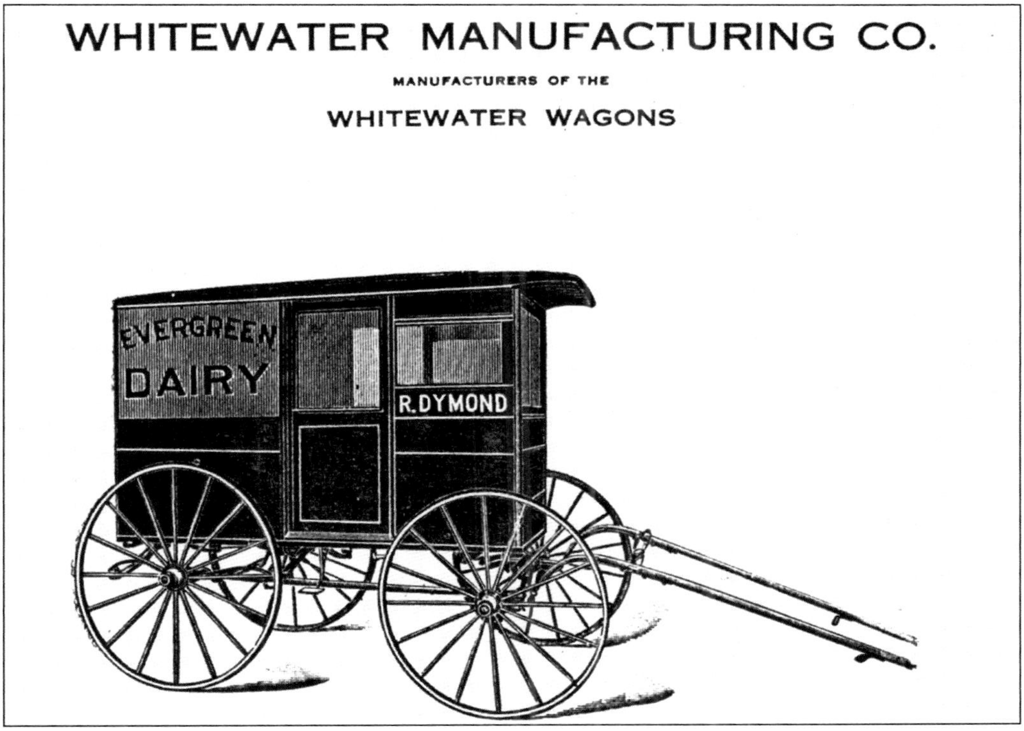

Weyher's Low-Down Wagon had a low floor between its wheels, close to ground level. Most wagons had their floors at the level of the axels or higher, so drivers of those wagons had to climb up to get aboard and step or leap down to get out. For door-to-door deliverymen, this repetitive process could be very tiring. Weyher's wagon attempted to solve this problem. (Irvin L. Young Memorial Library.)

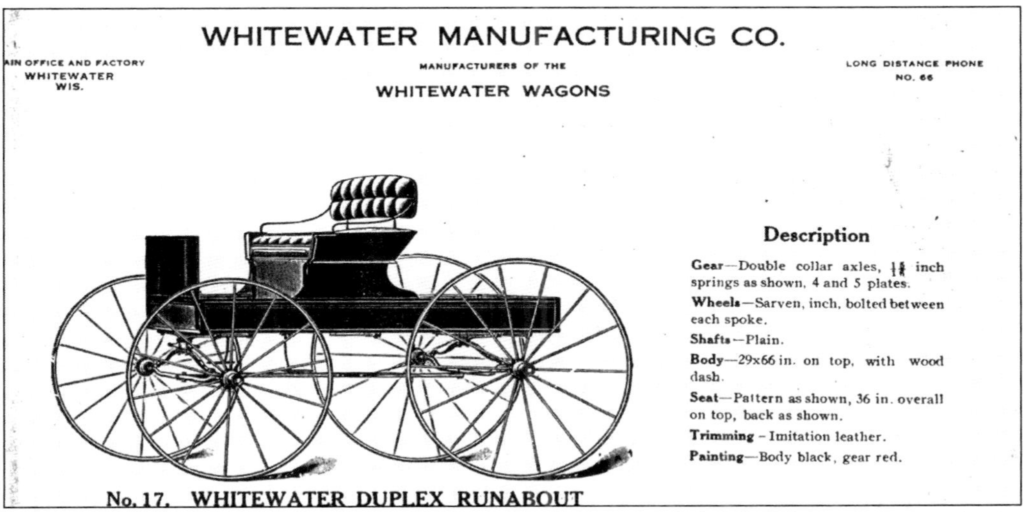

The Duplex Runabout, designed in the 1890s, was one of the last horse-drawn vehicles manufactured in Whitewater. It included such innovations as tubular steel running gear. However, the advent of the automobile hastened the decline of the Weyher Wagon plant. Faced with declining sales, Weyher failed in an attempt to manufacture his own motorized vehicle and was forced to shut down his factory in 1910. (Irvin L. Young Memorial Library.)

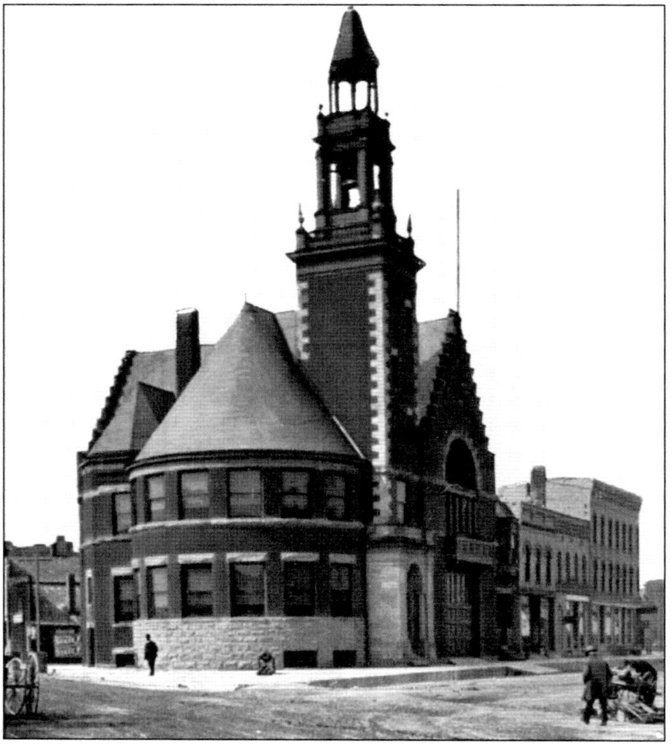

This photograph shows Whitewater's old city hall. The main floor housed the police department, city offices, and fire department equipment room. The municipal courtroom was on the second floor as well as a large meeting room where the council met. The bell tower served as the city's warning system. (WHS.)

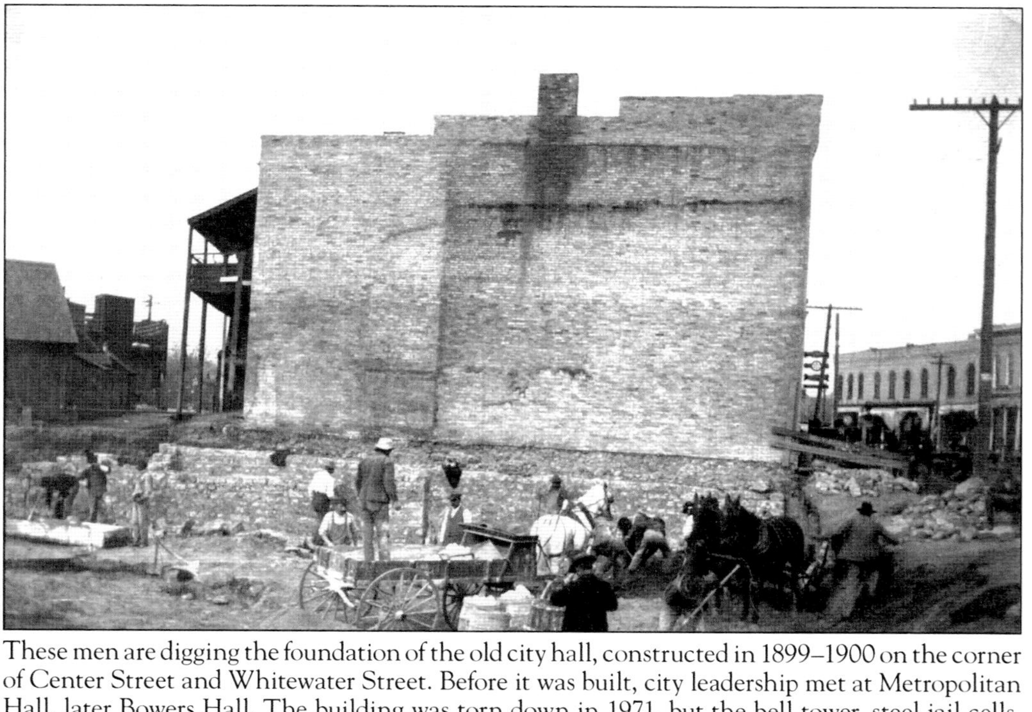

These men are digging the foundation of the old city hall, constructed in 1899–1900 on the corner of Center Street and Whitewater Street. Before it was built, city leadership met at Metropolitan Hall, later Bowers Hall. The building was torn down in 1971, but the bell tower, steel jail cells, and cornerstone were saved. (Irvin L. Young Memorial Library; Fran Achen Collection.)

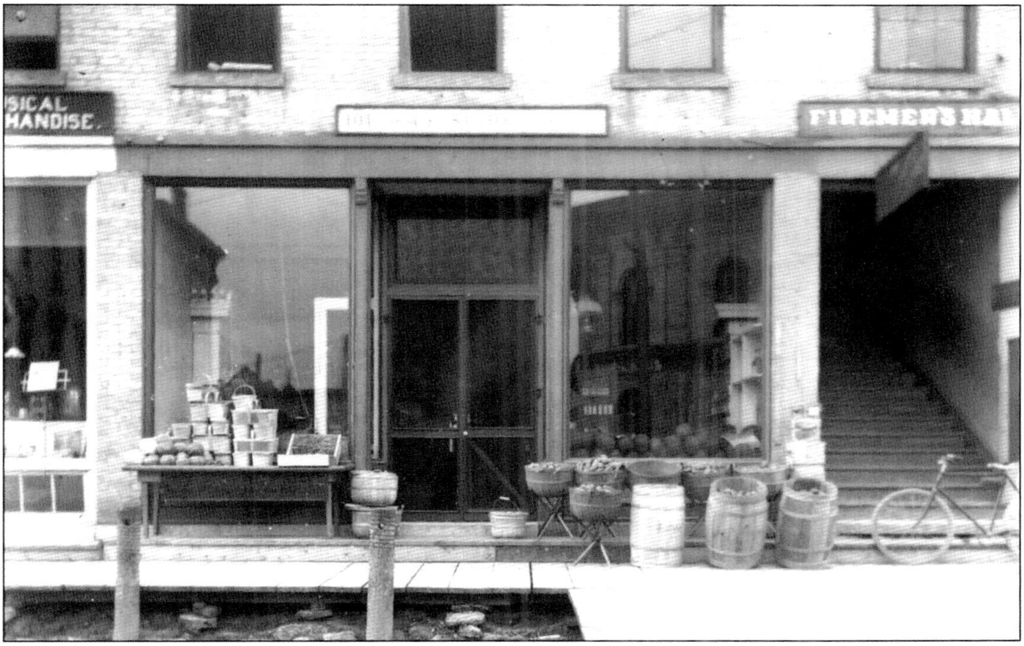

John Scholl opened his cooper shop in 1855 on Jefferson Street. Scholl produced mainly pork and whiskey barrels. Scholl's pork barrels were used in part at Whitewater's meatpacking factory, run by Sheldon C. Hall, and his whiskey barrels supplied Tuttle's distillery. His son Andrew opened A. E. Scholl's shop, pictured here on Center Street, which sold barrels and groceries. (WHS.)

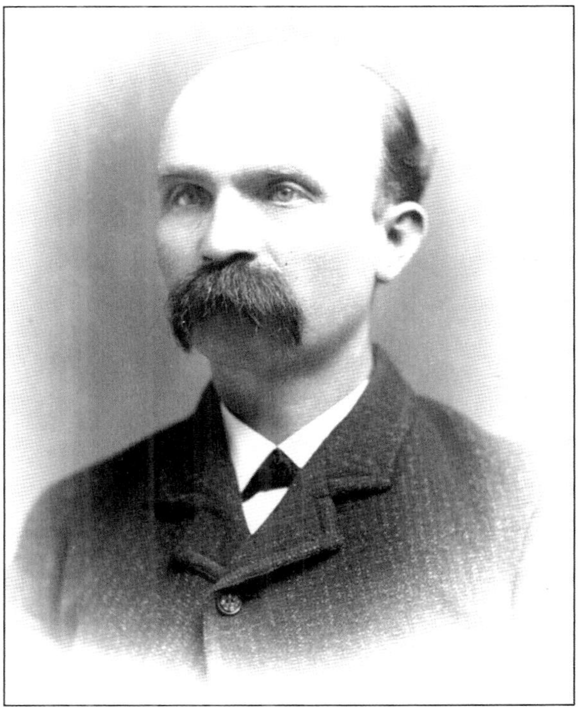

John Scholl's son Charles Scholl (pictured) joined the family cooper business and was instrumental in producing some of the last pork barrels made in Whitewater in 1886. The pork was salted, smoked, and packed into these large wooden barrels for storage and shipping, thus the term *meatpacking*. (WHS.)

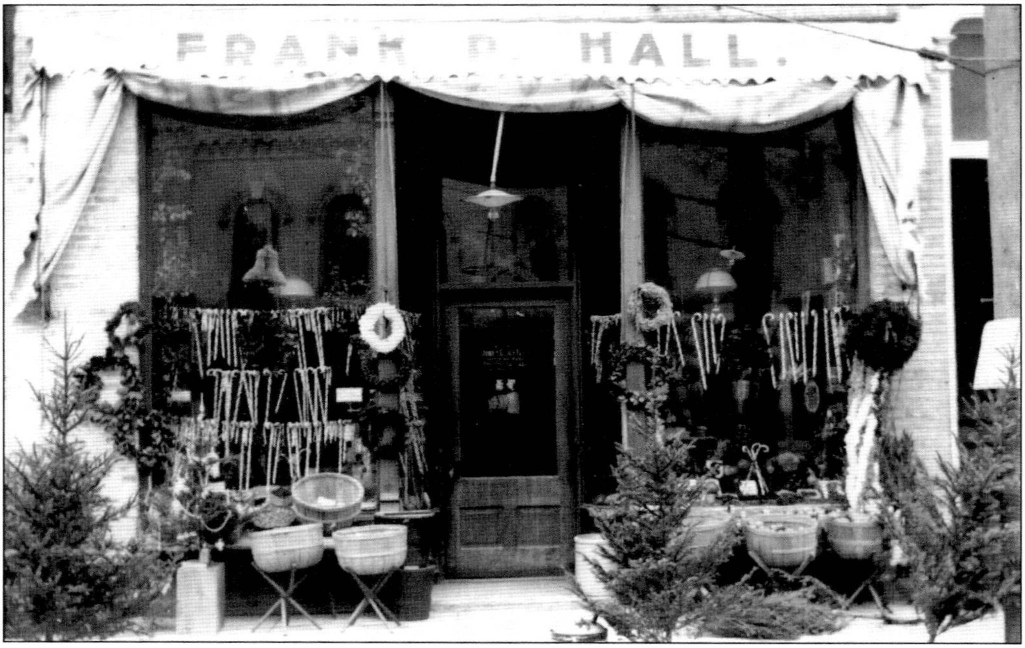

This popular picture of Frank P. Hall's grocery store at 106 Center Street shows Hall's Christmastime offerings displayed in front. Hall first entered the grocery business with John Partridge and Frank R. Hall, his uncle. He retired in 1917, selling his shop to Lawrence Bonney. The doorway is printed with "Jones Dairy Farm Little Pig Pork Country Sausages," first advertised in 1903. (Irvin L. Young Memorial Library; Fran Achen Collection.)

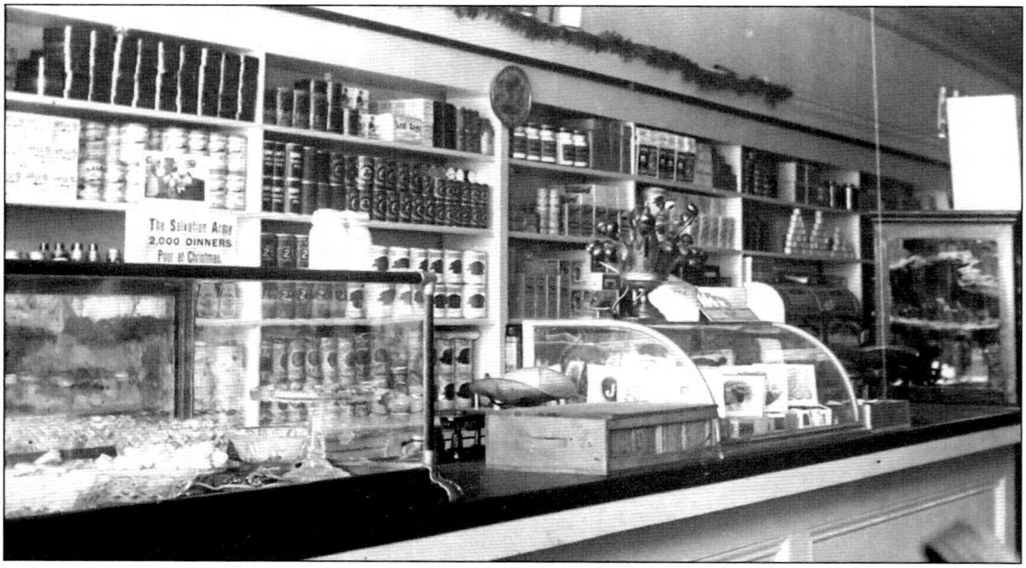

During the early 20th century, American grocery stores sought to make their shops cleaner. Frank P. Hall followed the trend, naming his shop the "F. P. Hall Sanitary Grocery." It was equipped with covers or glass cases to protect exposed foods at all times. The addition of large lights helped customers see better and gave a bright, clean appearance to the establishment. (Irvin L. Young Memorial Library; Fran Achen Collection.)

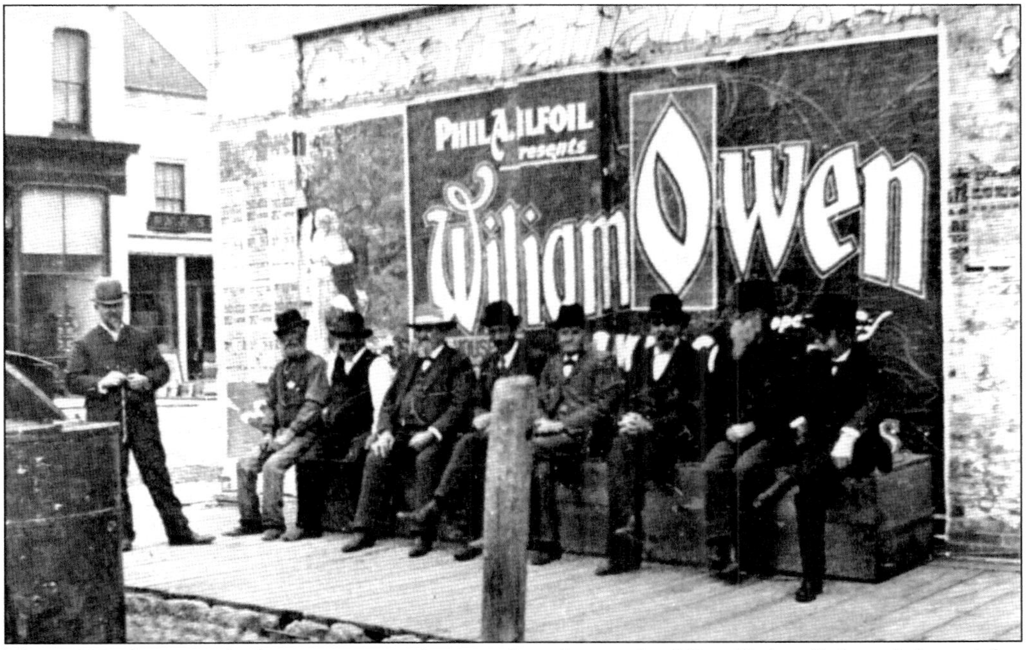

This 1899 photograph shows a group of men referred to as the "City Fathers": from left to right, John H. Fryer, postmaster; John Knight, pound master; Mike Messerschmidt, alderman and hotel proprietor; Lyman M. Goodhue, former mayor and owner of a furniture store; Walter Parish, landlord; Harvey Arveson, dry goods clerk; P. Dillon; Salmon H. Tuttle, weed commissioner; and Oscar Smith, carpenter. (Irvin L. Young Memorial Library; Fran Achen Collection.)

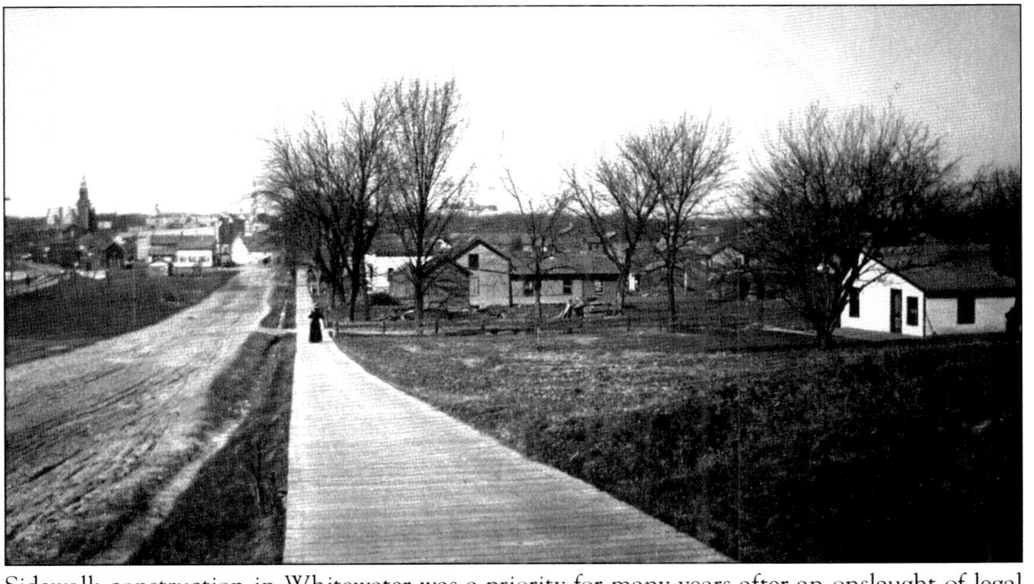

Sidewalk construction in Whitewater was a priority for many years after an onslaught of legal claims from citizens who were injured on dangerous walkways. However, this construction placed a serious strain on the city treasury. In 1893, Mayor Lyman M. Goodhue strongly recommended that "no new walks be built unless it is absolutely necessary to protect the city against claims growing out of defective walks." (Irvin L. Young Memorial Library; Fran Achen Collection.)

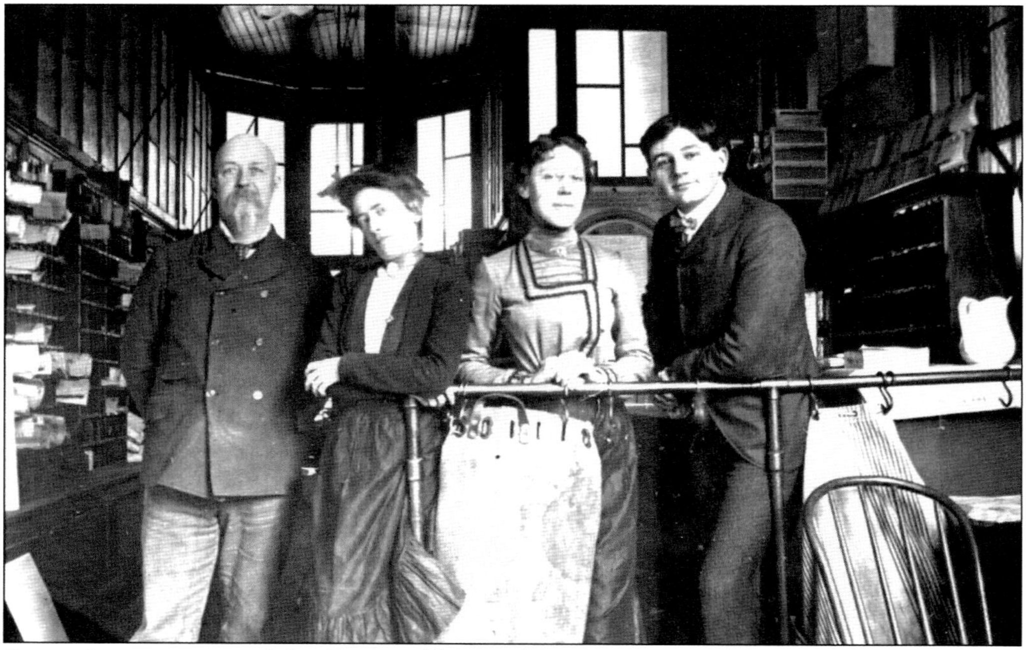

Pictured are postmaster John Fryer and his employees, Minnie Webb, Lena Hahn, and Henry Allen. The Whitewater Post Office was first located on First Street. It later moved to the corner of Main Street and Whitewater Street, where it remained until the new post office was built on Center Street in 1935. (Irvin L. Young Memorial Library; Fran Achen Collection.)

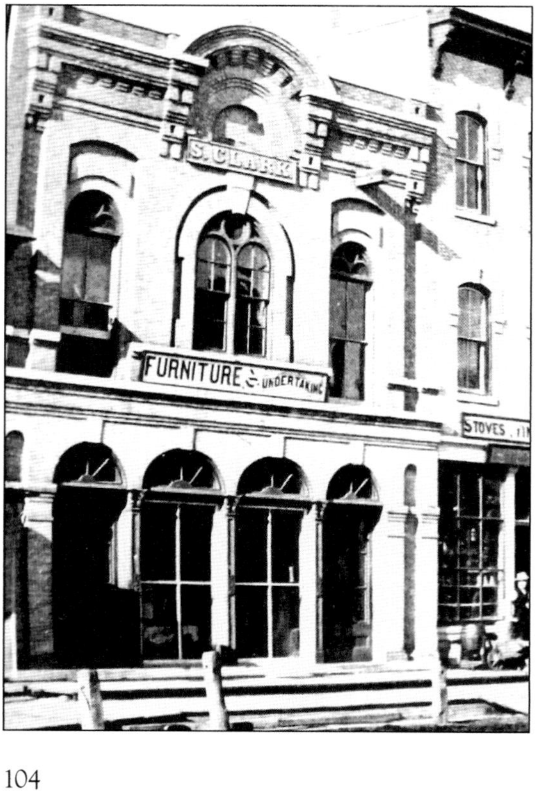

Salmon Clark came to Whitewater in 1850 and built his shop on Main Street near the corner of Second Street. Clark sold furniture and served as an undertaker. He also sold stoves, agricultural tools, and most kinds of hardware, copper, and sheet iron, as well as furnished goods to peddlers. Roswell Coburn purchased his shop in 1873. (WHS.)

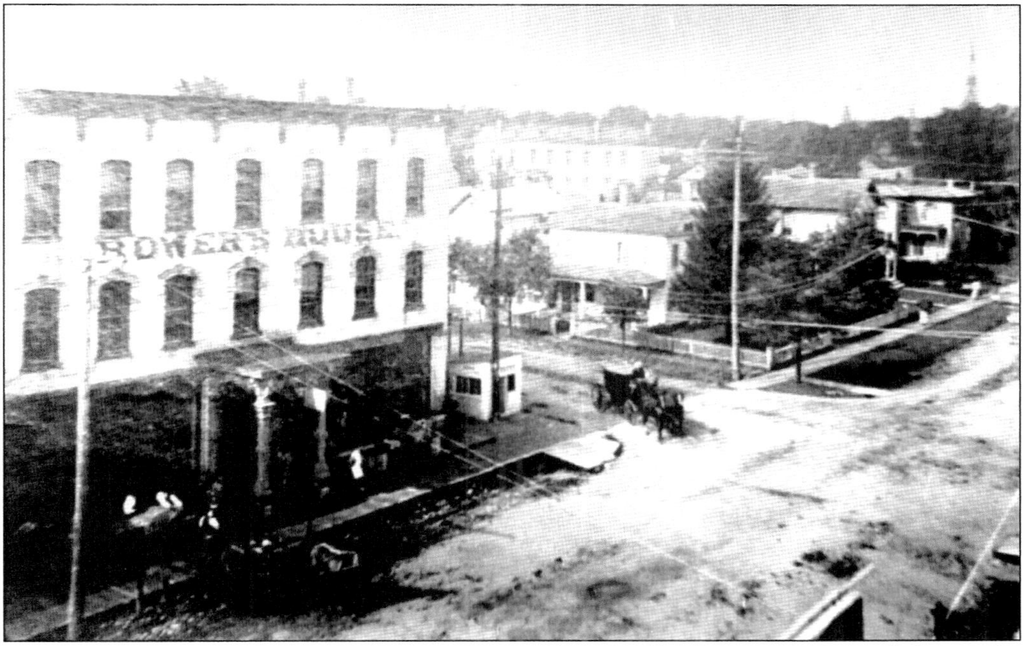

Joseph C. Bower came to Whitewater in 1846 to help build a brewery and decided to make Whitewater his home. Bower eventually became proprietor of both Bower's opera hall and the Bower House Hotel. The opera hall was built in 1870 with seating for 800 people. The 40-room hotel on Main Street was opened for business on September 9, 1880. Bower served for many years as the town treasurer. (Irvin L. Young Memorial Library; Fran Achen Collection.)

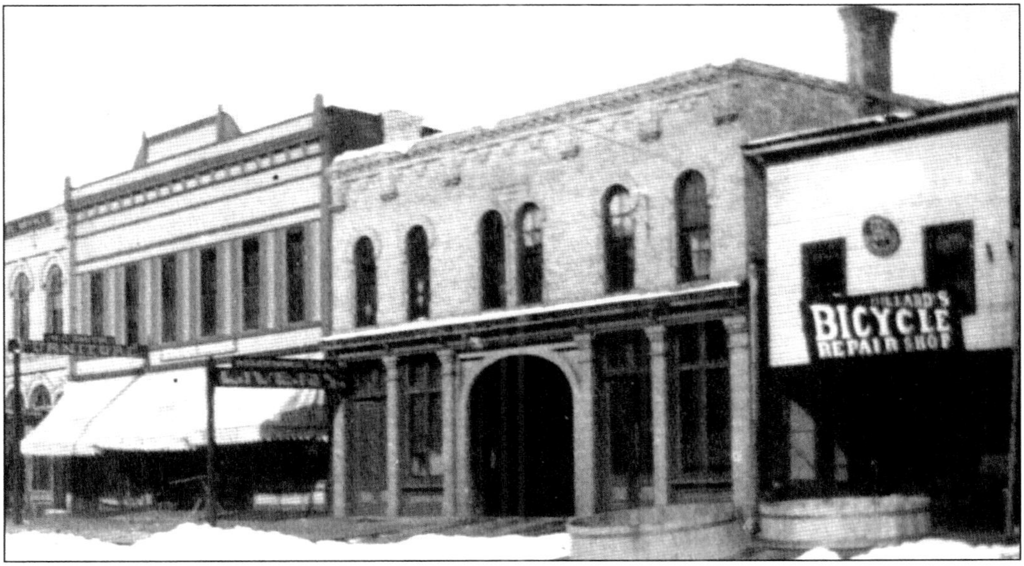

Leo Millard, a lifelong resident of Whitewater, ran the Millard Machine Shop and Bicycle Repair. He advertised as a patent worker and steam engineer. He also offered brass and foundry work and tool sharpening, in addition to repairing machines and bicycles. His shop, pictured here around 1900, was located on Main Street just west of the old stone mill. (Irvin L. Young Memorial Library; Fran Achen Collection.)

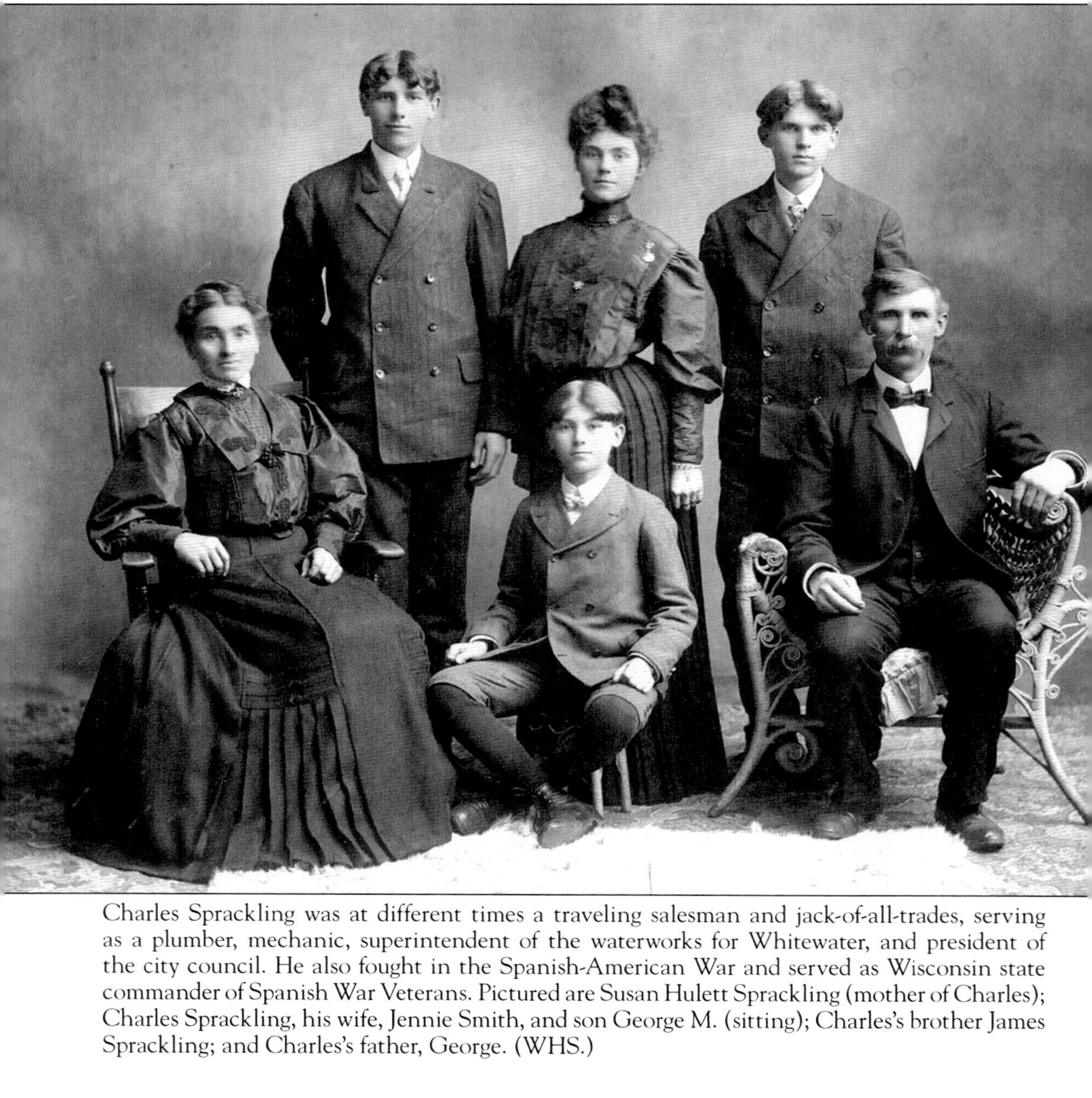

Charles Sprackling was at different times a traveling salesman and jack-of-all-trades, serving as a plumber, mechanic, superintendent of the waterworks for Whitewater, and president of the city council. He also fought in the Spanish-American War and served as Wisconsin state commander of Spanish War Veterans. Pictured are Susan Hulett Sprackling (mother of Charles); Charles Sprackling, his wife, Jennie Smith, and son George M. (sitting); Charles's brother James Sprackling; and Charles's father, George. (WHS.)

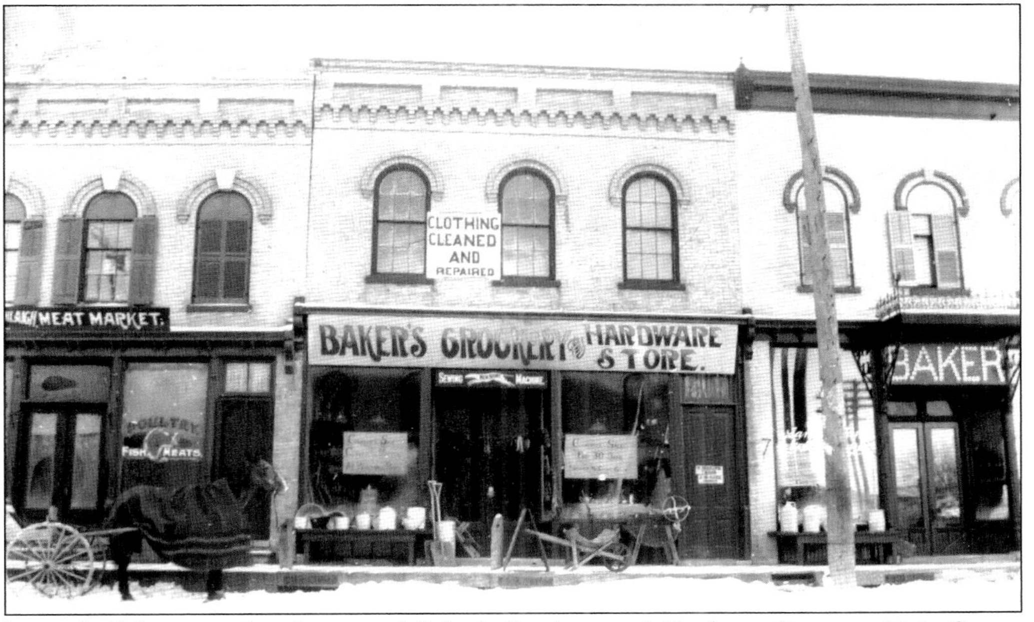

Louis C. Baker owned and operated Baker's Crockery and Hardware Store on Main Street. Baker started the business in 1895. By 1916, he had added W. H. Coxe as a partner. The store eventually occupied two buildings and offered a variety of goods, including hardware, fencing, cutlery, sewing machines, plumbing supplies, stoves, and bicycles. (Irvin L. Young Memorial Library; Fran Achen Collection.)

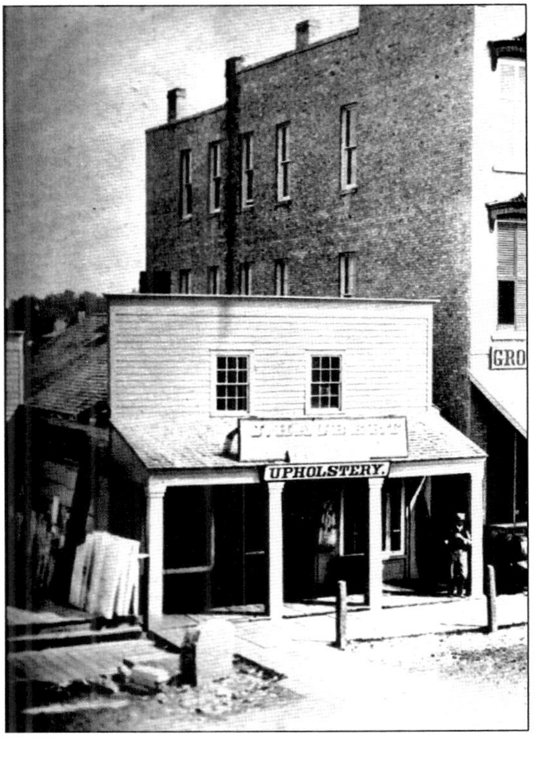

Joseph Haubert opened the Whitewater Harness Shop in 1862. Within a year, he had built a successful business that carried leather saddles, harnesses, trunks, valises, robes, blankets, whips, and all other types of "horse gear." He also offered upholstering services. The business was located on the north side of Main Street. Haubert's son Julius owned and operated a branch of the harness shop in Fort Atkinson. (WHS.)

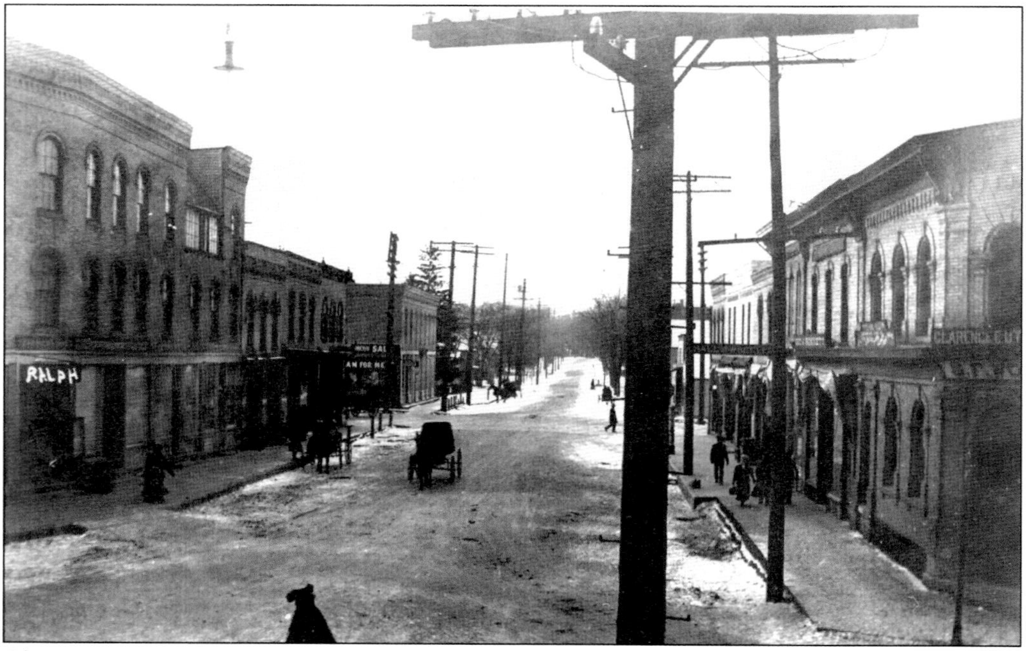

Clarence E. Utter opened his grocery business upstairs on Center Street and First Street in 1889. He sold teas, coffees, spices, fruits, canned goods, dairy, potted goods, confections, and cigars. He also dealt in farm produce, buying in bulk to sell at the lowest possible prices. In 1892, he moved his growing business to a larger building on Main Street. At left is the photography studio of William and Luella Ralph. (WHS.)

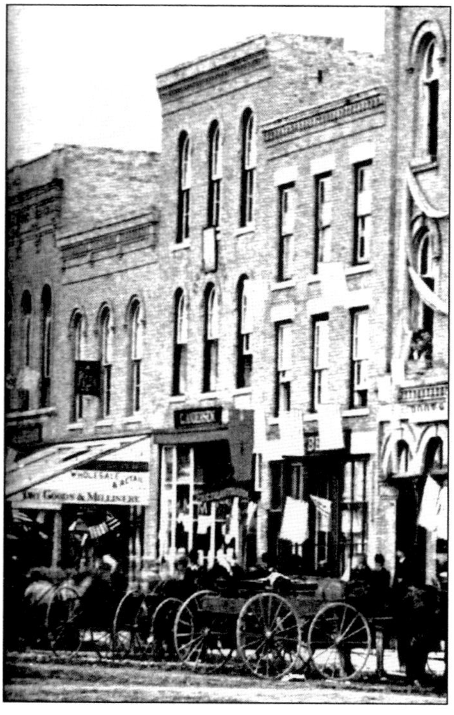

Gilbert Andersen came to Whitewater from Norway in 1853. By 1862, he had established a tailor business with Gullik Halverson as his partner. Andersen's store was located on Main Street. He had a custom tailoring business and offered ready-made clothing. He was also an agent for steamship lines to and from Europe. In addition, Andersen owned large interests in cattle raising and banking in Wyoming. (Irvin L. Young Memorial Library; Fran Achen Collection.)

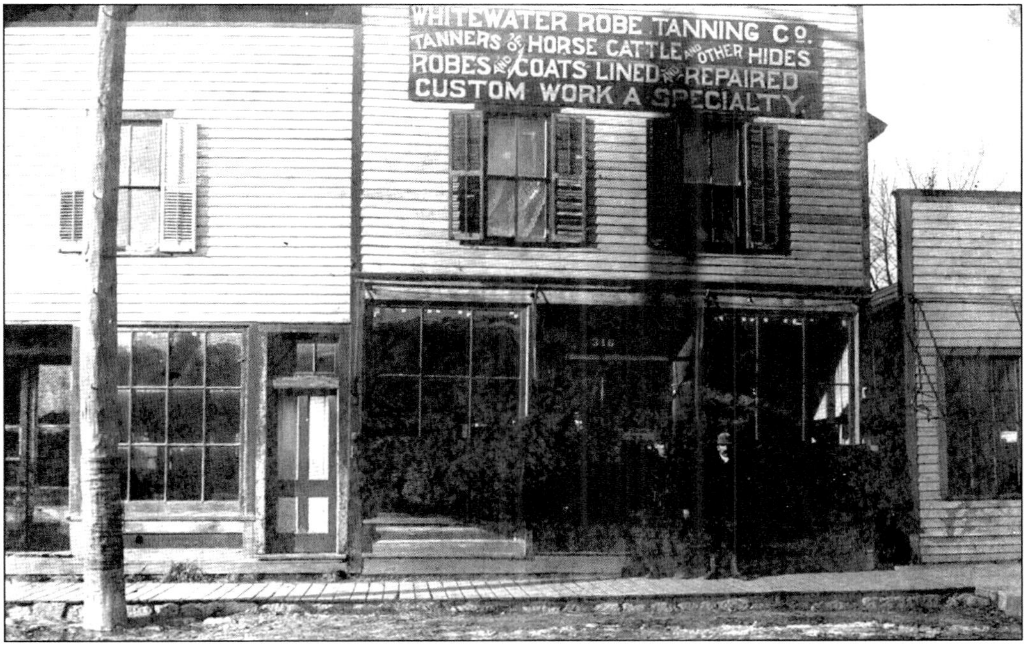

The Whitewater Robe Tanning Company was operated by Carl H. Ludtke. The building was located across from the railroad depot on Whitewater Street, between Fremont Street and Fourth Street. Ludtke tanned horse, cattle, and other skins and advertised rugs, mittens, and ladies' garments. Animal skin robes were most commonly used as blankets for riding in open carriages. (Russ Fero.)

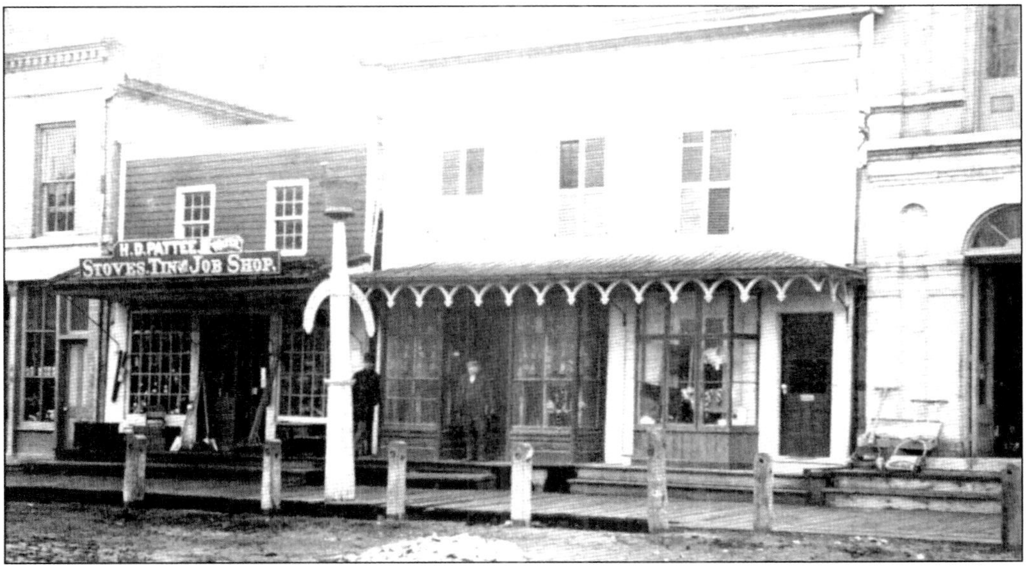

Dr. John Deichman came to Whitewater in 1849 after studying medicine at the University of Pennsylvania. Deichman operated the Whitewater Drug and Chemical Store, center, and served as a druggist, apothecary, and taxidermist. He sold drugs, paints, liquors, tobacco, fishing tackle, toys, trusses, and confectionary. He also had a large collection of parrots and fish, as well as many stuffed birds, squirrels, weasels, badgers, snakes, frogs, and lizards. (WHS.)

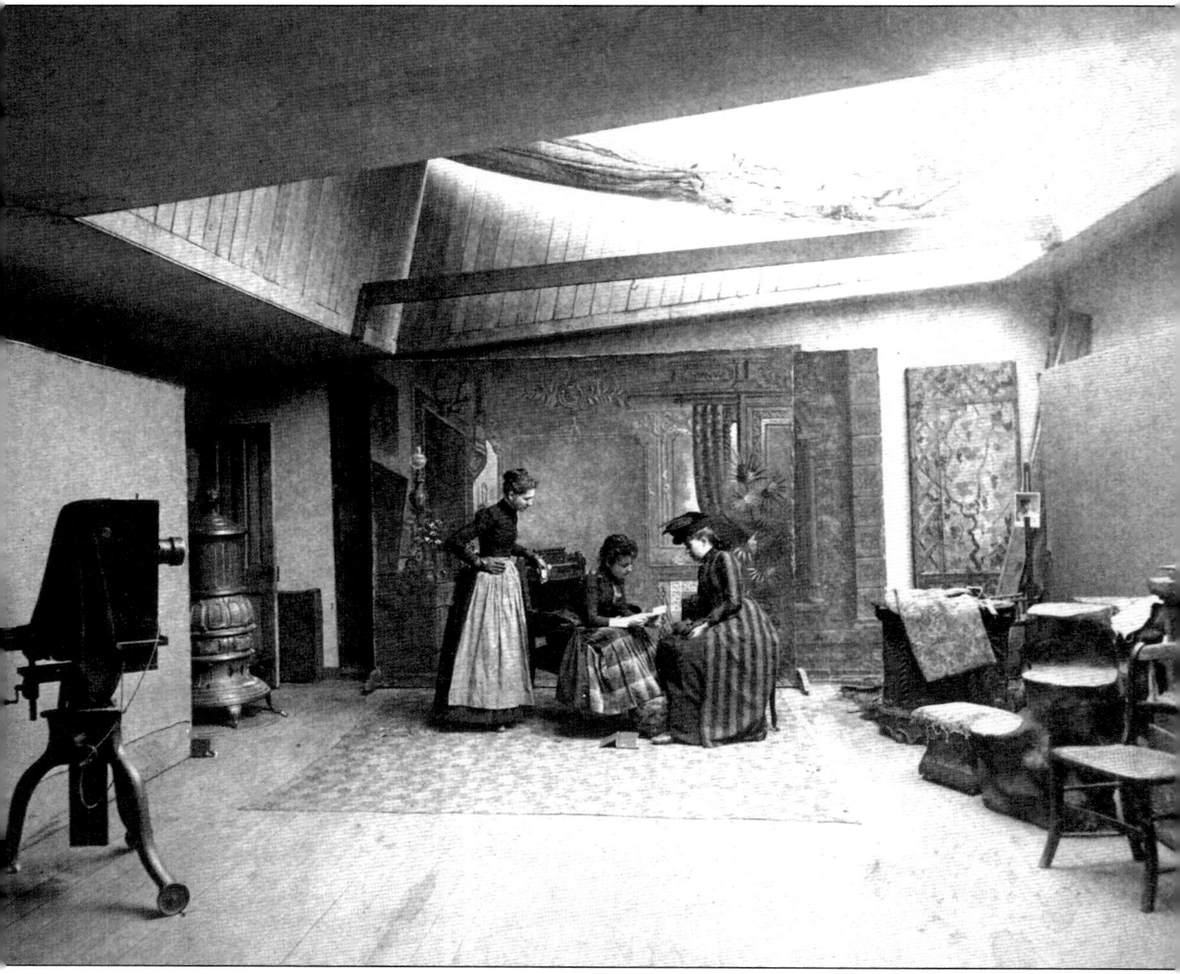

Harlan P. Goodman, photographer, opened for business in 1878 and continued for more than 30 years. At the time of his opening in 1878, he advertised card photographs for a mere $1.50 per dozen. Pictured in Goodman's original studio are Goodman's daughter Lottie, Jennie Smith, and Cora Towle. Goodman's services included copying and enlarging images. He also colored photographs with oil, crayon, ink, and watercolors. (WHS.)

Six

The Dairy Era

What was to become of the town, then, if its factories no longer remained? Dairy farming was not as glamorous as manufacturing or as easy as wheat farming, but Wisconsin as a whole was in the middle of the great shift "from the plow to the cow." Therefore, Whitewater embraced its potential as a dairy town. Perhaps this shift was made easier by dairy's relative stability during the 1890s. Legendary Wisconsin dairyman William Dempster Hoard said that dairy regions "met the flood of disaster with a smiling front and without a tremor."

The dairy industry was growing, and Whitewater was caught up in this growth, as cheese making and creameries helped carry Whitewater into the 20th century. Ultimately, it was Whitewater's ability to adapt to a changing agricultural climate that allowed it to remain a stable, albeit much smaller, agricultural town. In 1871, Whitewater had two large cheese factories. By the close of the 19th century, southeastern Wisconsin's American cheese production was concentrated in Whitewater. According to a state labor report from 1906, Whitewater was in the midst of "an extensive cheese and butter manufacturing district which has resulted in being a shipping point for immense quantities of butter, cheese, and eggs." Creameries and dairy farms also did much to help Whitewater's recovery, although the process was painstakingly slow. Not until the 1940s did Whitewater's population climb back to the level the town had reached in 1890.

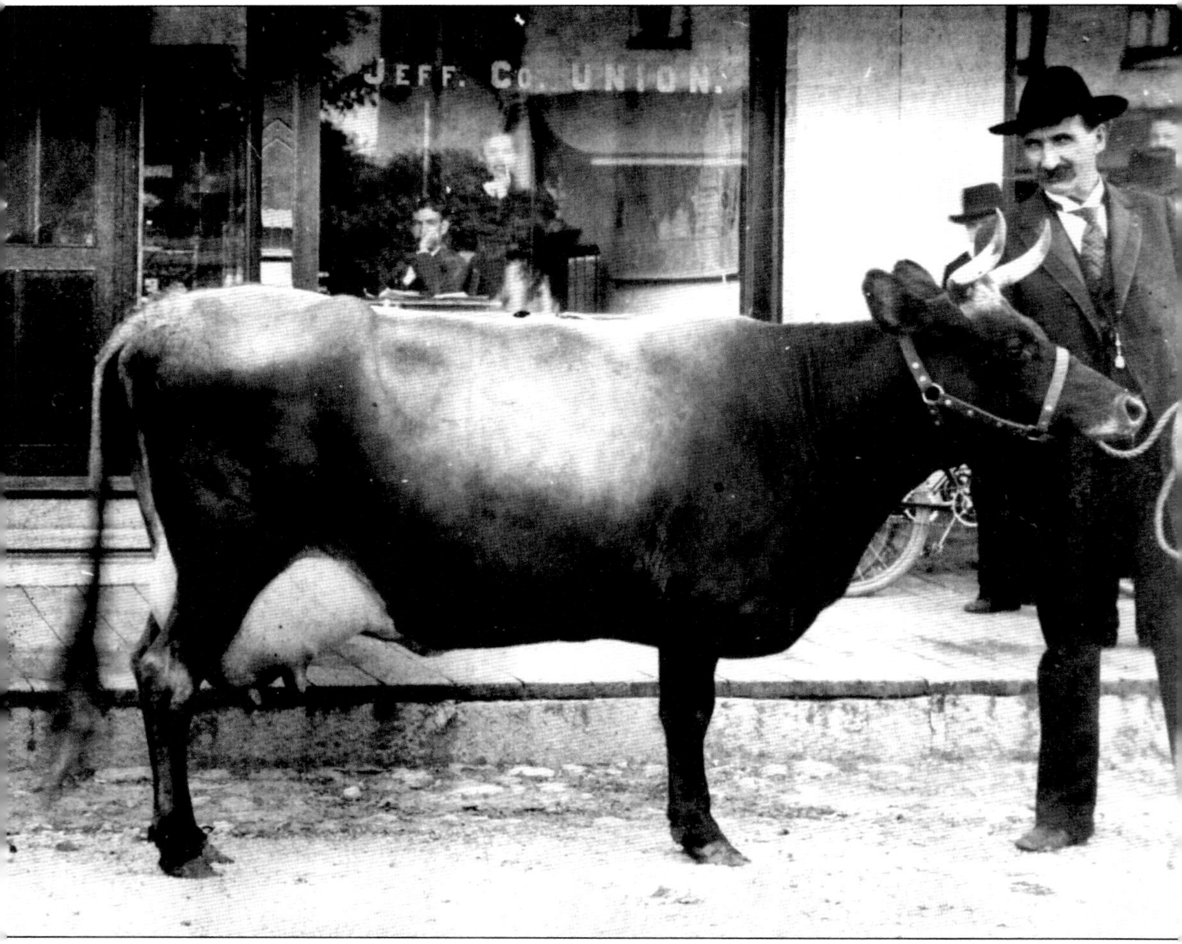

William Dempster Hoard, legendary dairyman from Fort Atkinson, is pictured here with a jersey cow. Hoard became president of the Northwestern Dairymen's Association in 1872. The organization began mainly in response to the failure of wheat crops, and Hoard argued that dairy was more consistent and predictable than crop growing. Hoard also created *Hoard's Dairyman*, the national dairy farm magazine, in 1885. (Hoard's Dairyman.)

The Wisconsin Dairy Supply Company was established in 1883 as a partnership between John C. Kachel and Thomas A. Kachel. It was the largest dairy supply company in the state by 1916. In 1900, the company's growth prompted a move to the location of the old Winchester and Partridge wagon factory. The company manufactured supplies and equipment for cheese factories and creameries. (Irvin L. Young Memorial Library; Fran Achen Collection.)

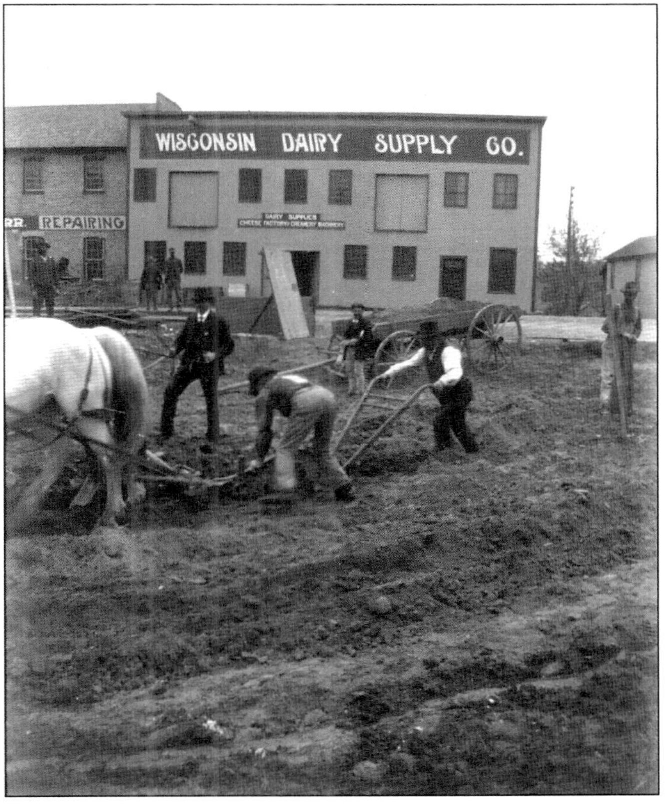

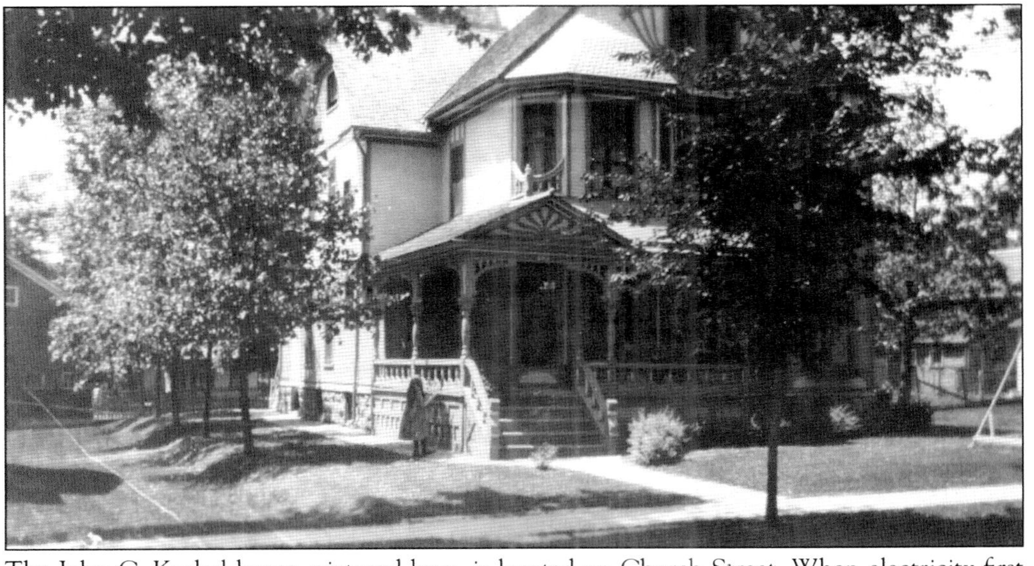

The John C. Kachel house, pictured here, is located on Church Street. When electricity first came to Whitewater in the 1890s, an electrical line was built to service the lumberyards on Whitewater Street, and the Kachel house was situated along the line. The Kachel family connected their home to the electrical line, making their house the first in Whitewater to have electrical power. (WHS.)

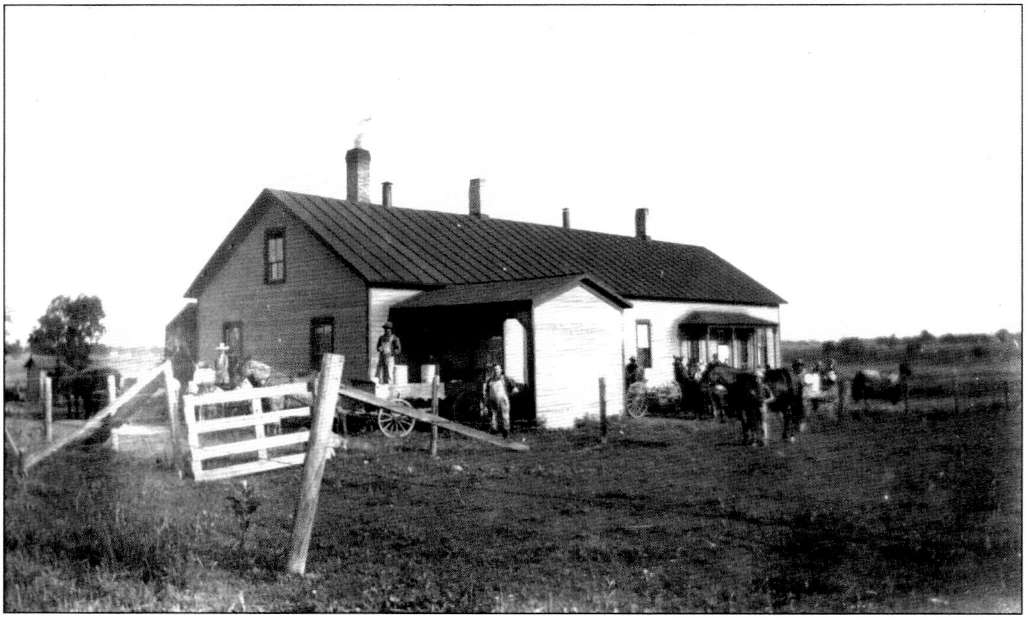

The Island Cheese Factory was operated by an Englishman named William Stowe who lived in LaGrange Township. According to the Wisconsin Dairyman's Association, in 1876, Stowe produced 64,000 pounds of cheese. The Island Cheese Factory was located on Island Road in Richmond, just south of Whitewater. The factory burned down in 1890. (Irvin L. Young Memorial Library; Fran Achen Collection.)

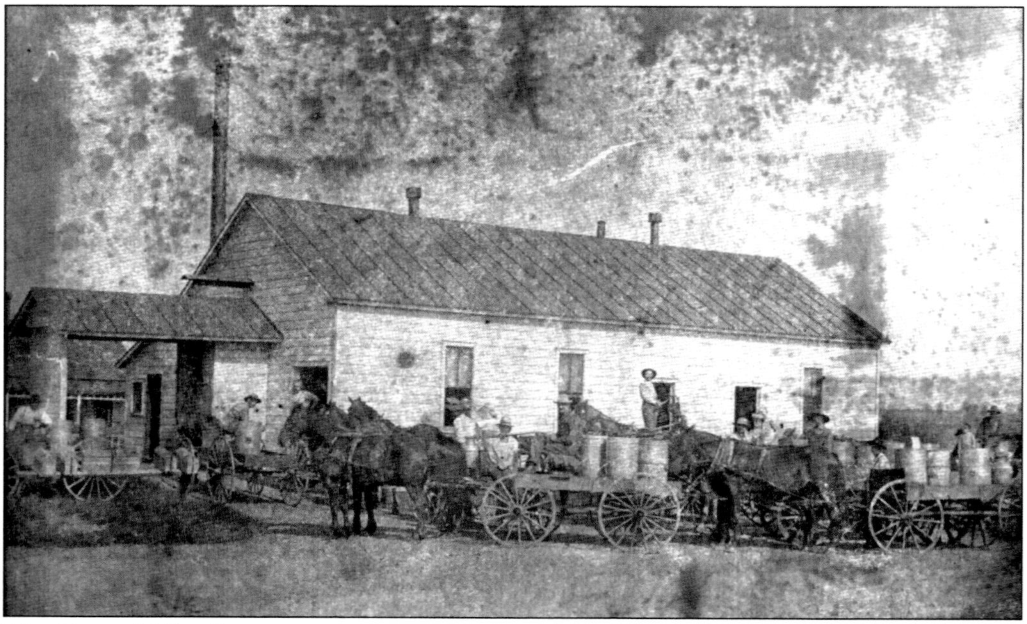

The Clover Valley Creamery, pictured around 1905 in Richmond, was owned and operated by George Fryer and produced mainly butter. Fryer came from Sharon after working in a cheese factory and returned by 1910 to Sharon to work in a creamery. Fryer purchased the creamery from William Wright and George McIntyre. (WHS.)

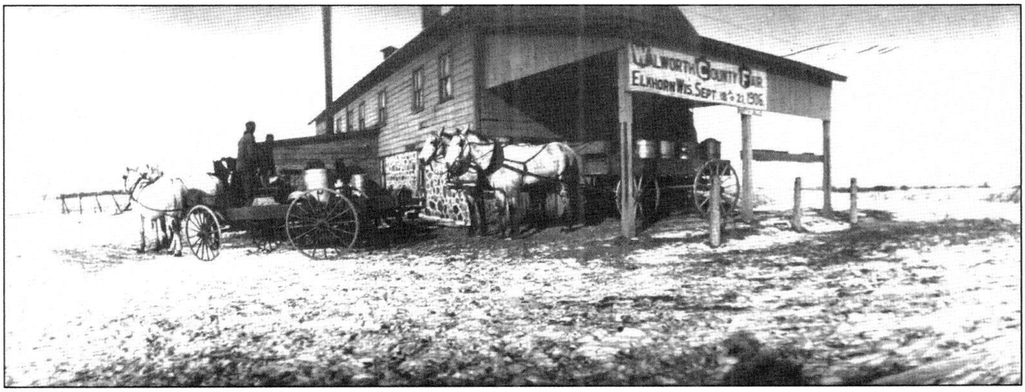

The Town Line Creamery, originally called the Whitewater Creamery, was built in 1890 by Frank Newell, who first intended to use the building as a cheese factory. It was located on the corner of Whitewater-Richmond Town Line and Highway 89. James Stewart and John Chapman later purchased the building and turned it into a creamery, which closed in 1917. The Town Line Creamery was later remodeled into the Whitewater Lake Inn. (WHS.)

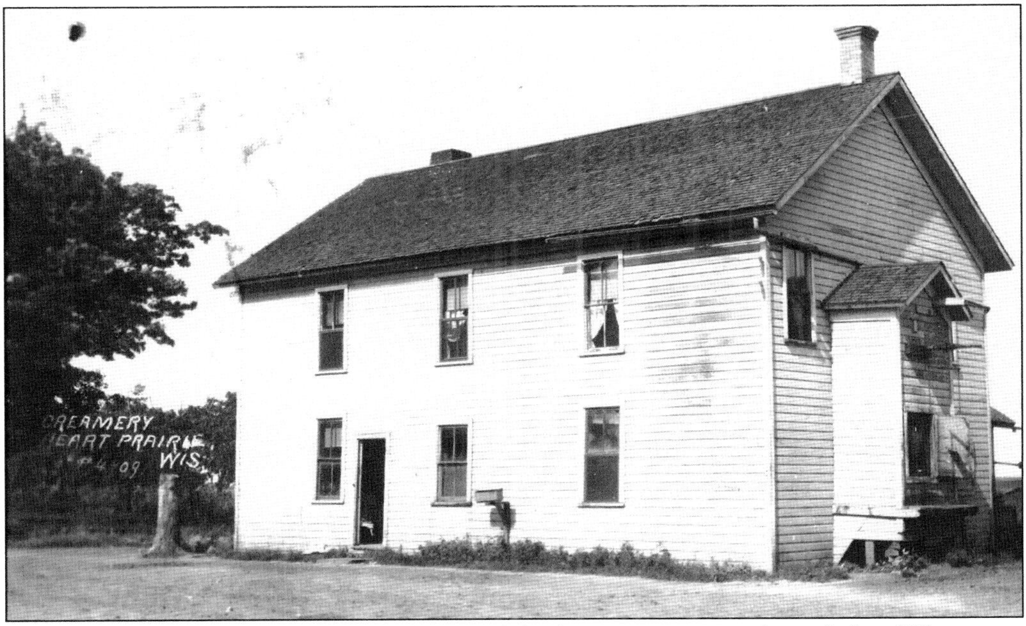

In 1902, the biennial report of the dairy and food commissioner indicated that Whitewater featured 16 creameries, including the Prairie Creamery, Nonpareil Creamery, Reliance Creamery, Clover Valley Creamery, Spring Prairie Creamery, and Eureka Creamery Company. These creameries were visited by state inspectors on a regular basis to make sure that their products were not skimmed or watered down. Pictured is Lewis Peterson's creamery in nearby Heart Prairie. (WHS.)

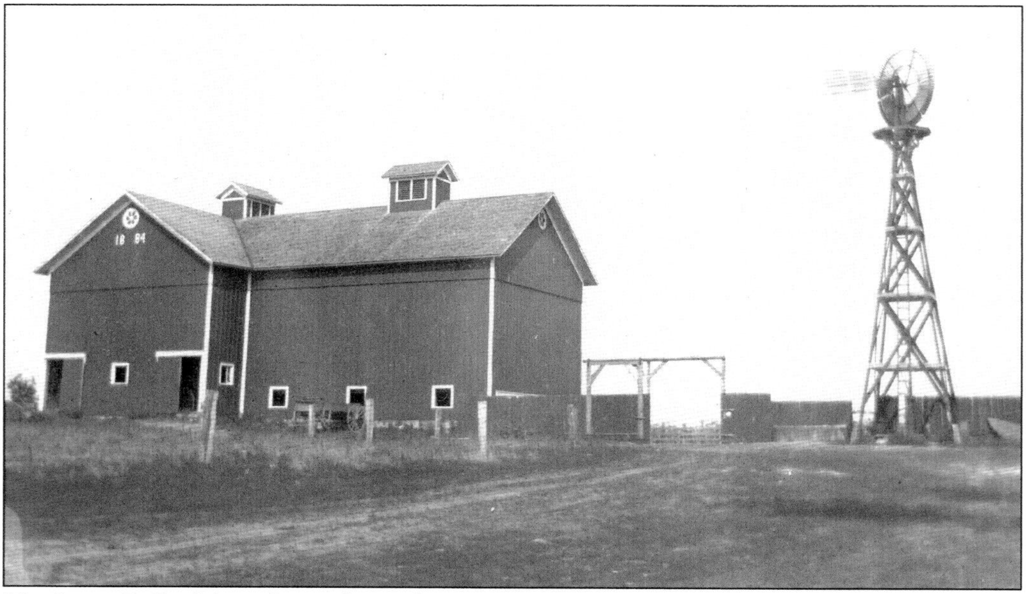

The James Pollard farm, located in Whitewater Township, is pictured here around 1895. After Pollard's death in 1892, his wife Elizabeth and daughter Belle remained on the farm. The Pollard farm produced some of the lowest-fat milk in Whitewater, with a 3.6-percent fat content; however, higher fat count was more desirable. The barn in the photograph is marked with "1884," the year of its construction. (Irvin L. Young Memorial Library; Fran Achen Collection.)

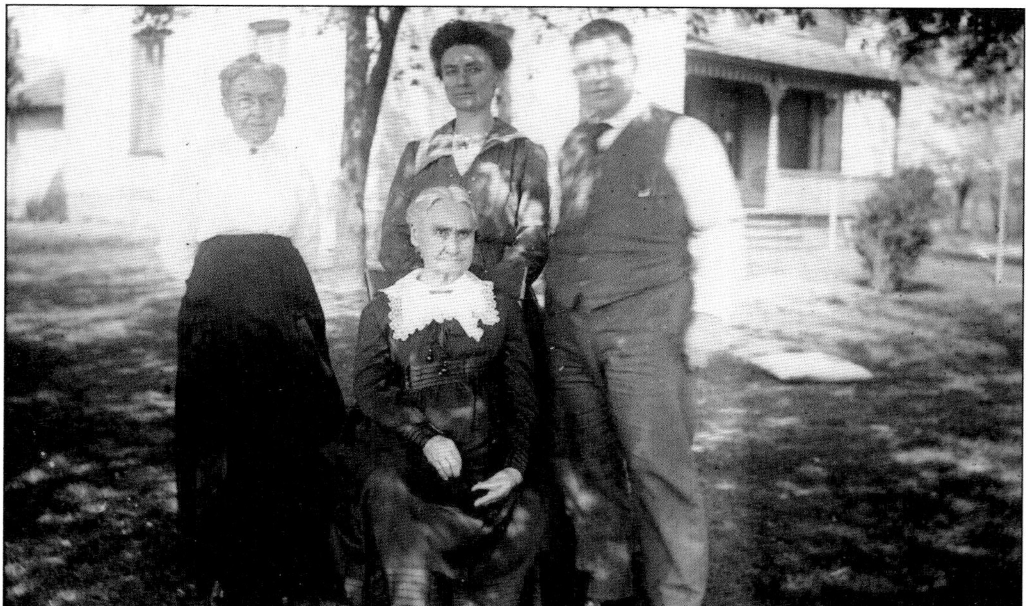

Pictured are Julia Hackett, Belle Pollard Scholl, Henry Scholl, and Elizabeth Hackett Pollard (seated). The Hackett farm was situated next to the Pollard farm in Whitewater Township. Belle, the daughter of James and Elizabeth Pollard, married Henry Scholl in 1907, and Belle, Henry, and Elizabeth soon moved into Whitewater to a home on Center Street near the Methodist church. (Irvin L. Young Memorial Library.)

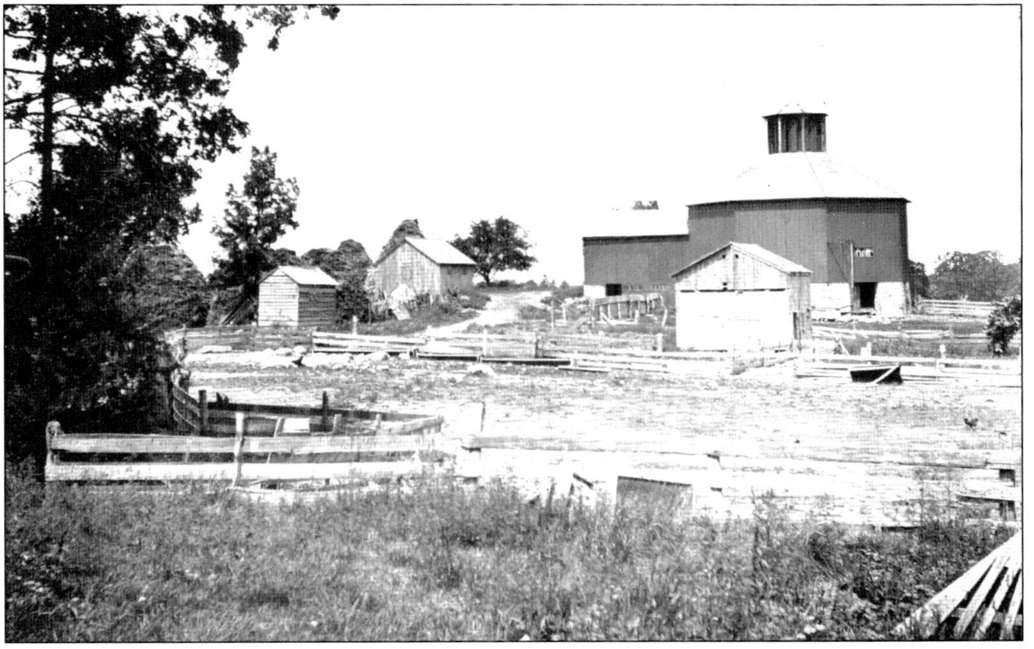

The Pollard/Scholl farm boasted one of the few octagonal barns in the area. The barns offered a more efficient use of space for dairy farming, but the building of such a barn involved an elaborate design with rooms in the shape of pie slices. To complicate the process, it was often the case that a farmer's neighbors would lack the architectural expertise to help construct such a barn. (Irvin L. Young Memorial Library; Fran Achen Collection.)

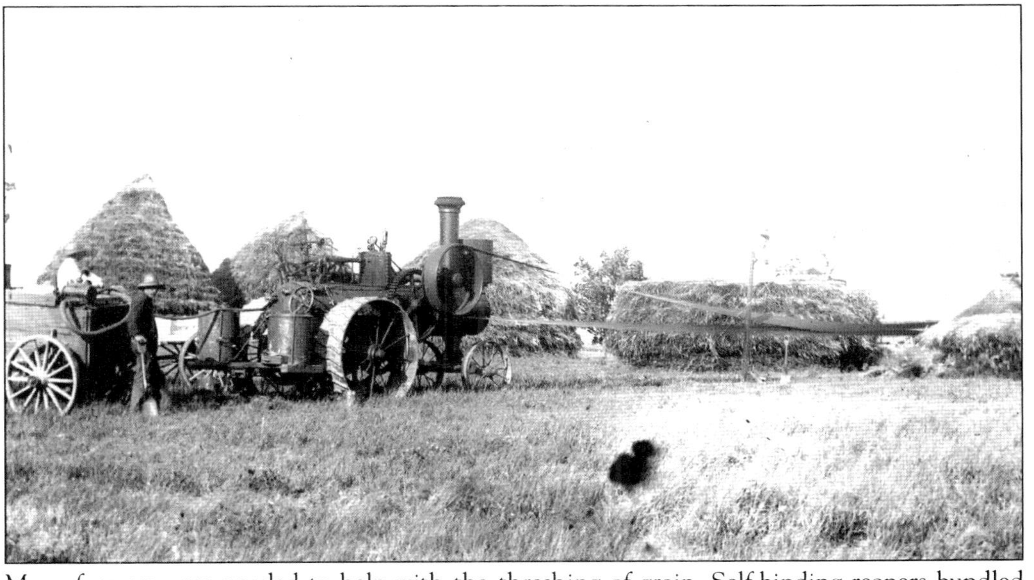

Many farmers were needed to help with the threshing of grain. Self-binding reapers bundled stacks of grain. These stacks were then manually set up in conical shocks, with the grain ends toward the center, to ensure runoff of precipitation. This c. 1895 photograph of the Pollard/Scholl farm shows an early tractor near many large stacks of grain. (Irvin L. Young Memorial Library; Fran Achen Collection.)

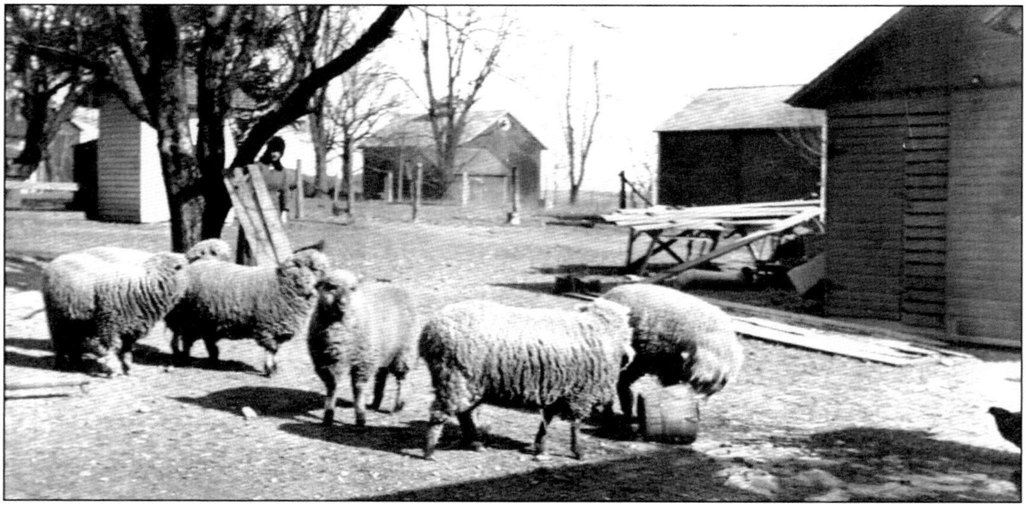

A. F. Knox, Leonard Smith, Herman Hemenway, Charles Clark, and John Clark began to import choice Merino sheep from Vermont in the early 1860s. The fine wool brought good prices, and Whitewater, for a while, became a center for sheep raising and a growing wool market. By 1862 Knox and Smith were able to sell their sheep for $30 a head, while cows were selling for only $15–$25 a head. By 1864, a farmer could expect nearly $18 for the wool of a single sheep. (WHS.)

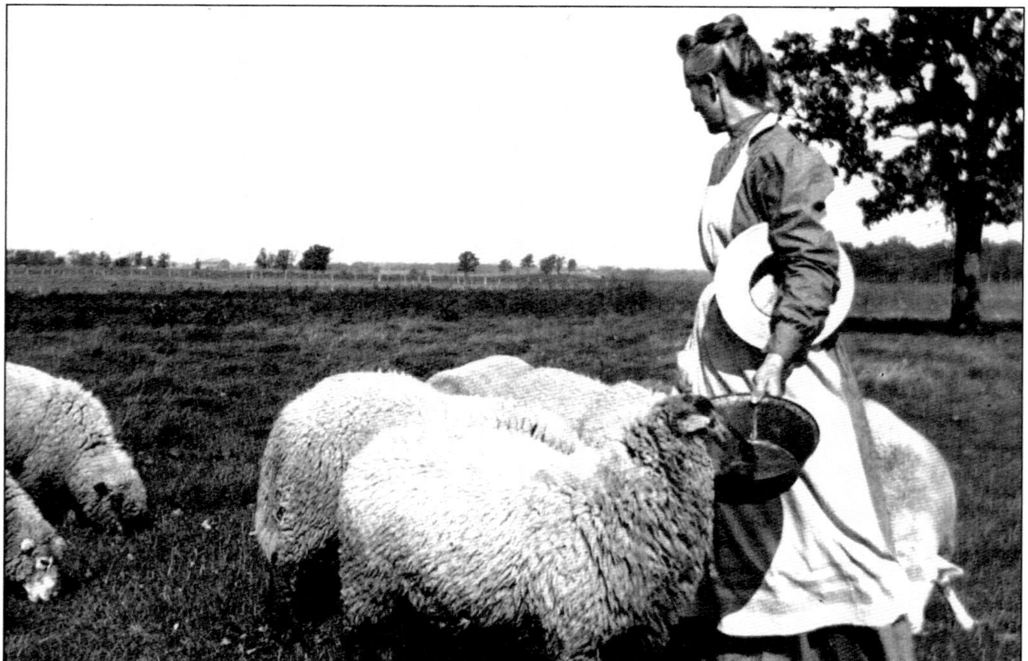

Nellie O'Neil, daughter of Thomas O'Neil, tended sheep on the family farm in Whitewater Township. Sheep raising became important in some southern Wisconsin counties when the Civil War created an incredible demand for wool to make hats and uniforms. By the mid-1870s, demand was much lower, wool prices dropped, and many sheep producers either reduced or eliminated their herds. (Irvin L. Young Memorial Library; Fran Achen Collection.)

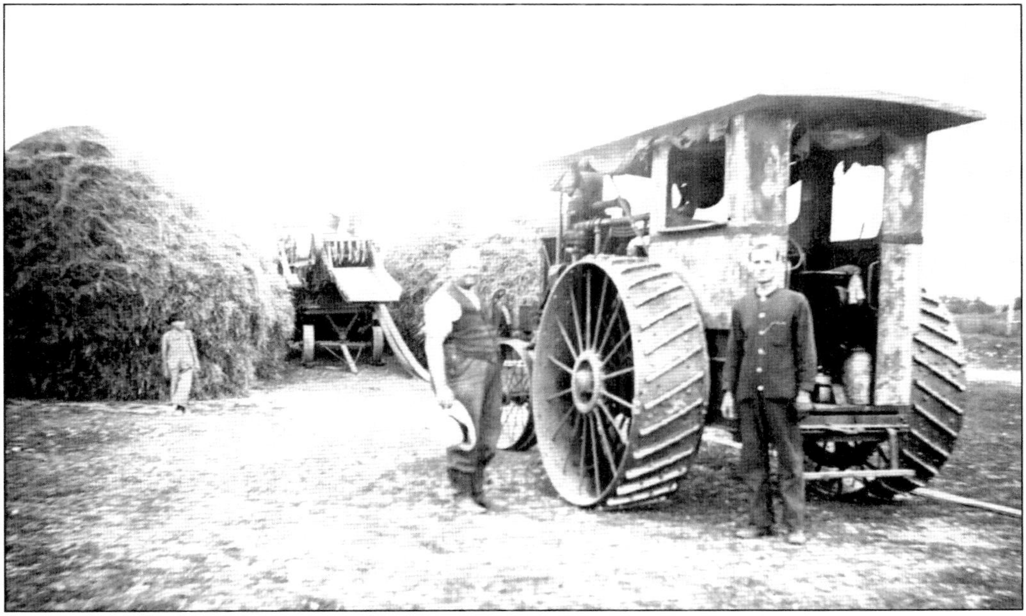

The William Bromley family came to Wisconsin in 1845. Their farm, pictured, was handed down to William's son George, who farmed in LaGrange until his death in 1926. Pictured is George's nephew Charlie Bromley, left of the tractor. The tractor is a 20/40 Case tractor, produced in Racine from 1912 to 1916. (Irvin L. Young Memorial Library; Fran Achen Collection.)

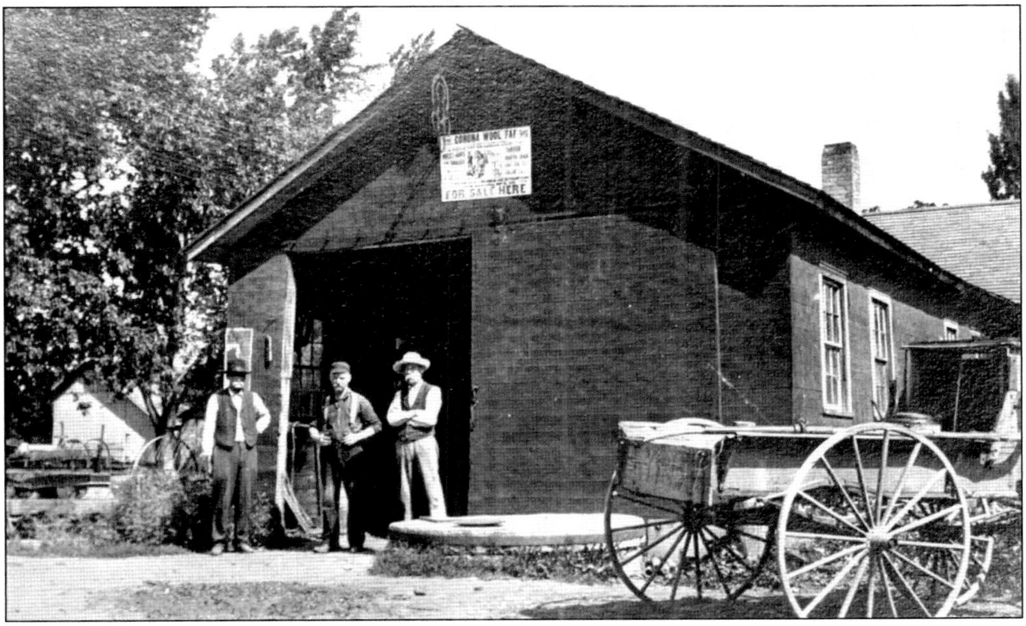

These men stand outside an unidentified Whitewater blacksmith shop around 1900. The shop advertises Corona Wool Fat, a salve made by the Corona Manufacturing Company of Kenton, Ohio. The lanolin-based compound was a healing ointment used on horses and cows to treat barbwire cuts, thrush, sore teats, and hard feet. (Irvin L. Young Memorial Library; Fran Achen Collection.)

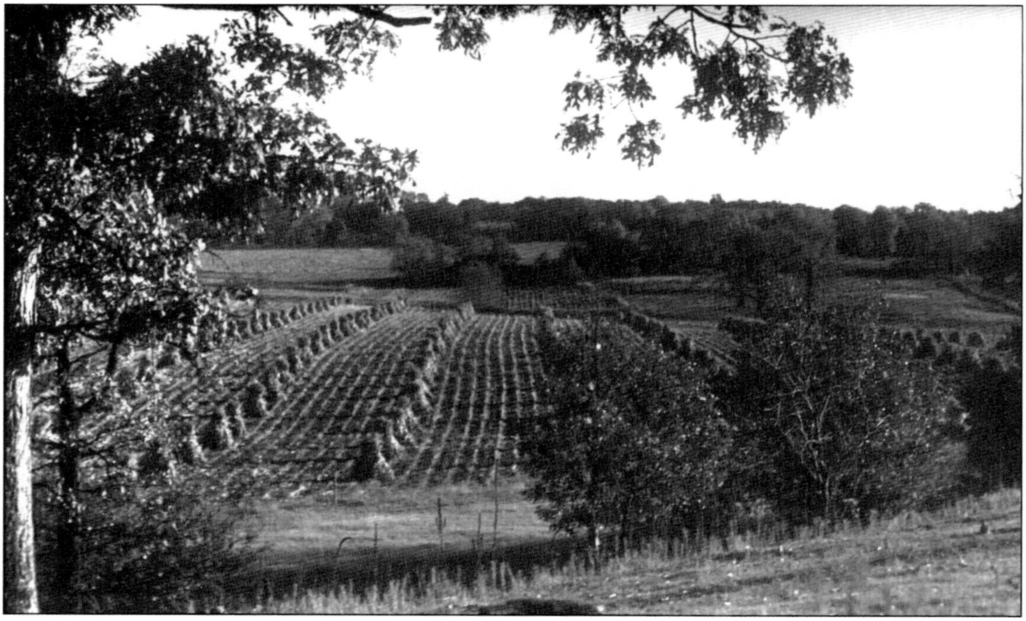

As the wheat frontier moved westward, Wisconsin farmers grew a greater variety of crops. Mid-19th-century farmers thought that Wisconsin would never be a corn state, but corn became a chief cash crop by 1900. New techniques and procedures, including a sophisticated crop rotation system, allowed farms to grow more corn in more places throughout the state. (Irvin L. Young Memorial Library; Fran Achen Collection.)

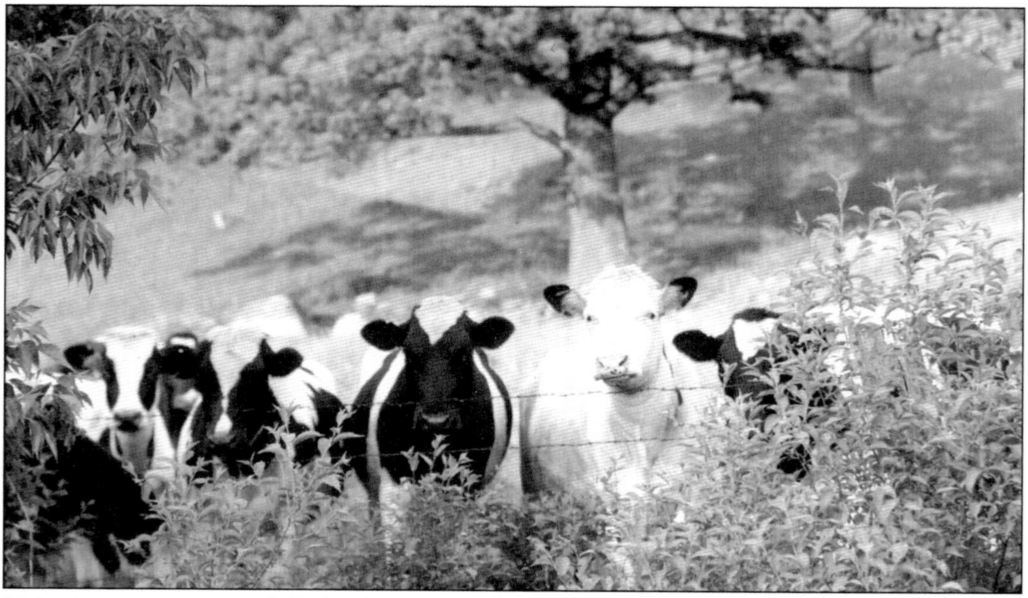

Holstein cows were carefully bred to create the most productive animal possible. These cows produced more milk and beef than any other breed, and Americans took serious interest in the Holstein during the mid-1800s. In 1850, there were 64,000 cows of all breeds in Wisconsin, but by 1905 there were more than 1.2 million cows in the state. (Irvin L. Young Memorial Library; Fran Achen Collection.)

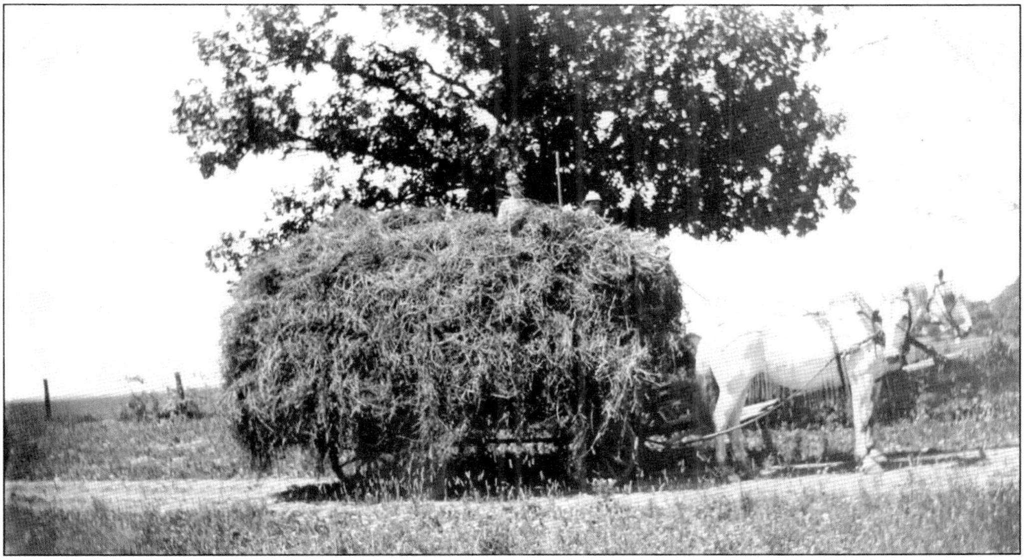

As dairy farming spread throughout Wisconsin, farmers had to provide more food to their herds. While oats were considered "the finest grain for producing milk that there is," hay was much cheaper and more commonly used. Hay production in Wisconsin increased from 275,000 tons in 1849 to 1,907,000 tons in 1879. Most farmers grew their own hay for feed, supplementing with alfalfa, corn meal, bran, and other grains. (Irvin L. Young Memorial Library; Fran Achen Collection.)

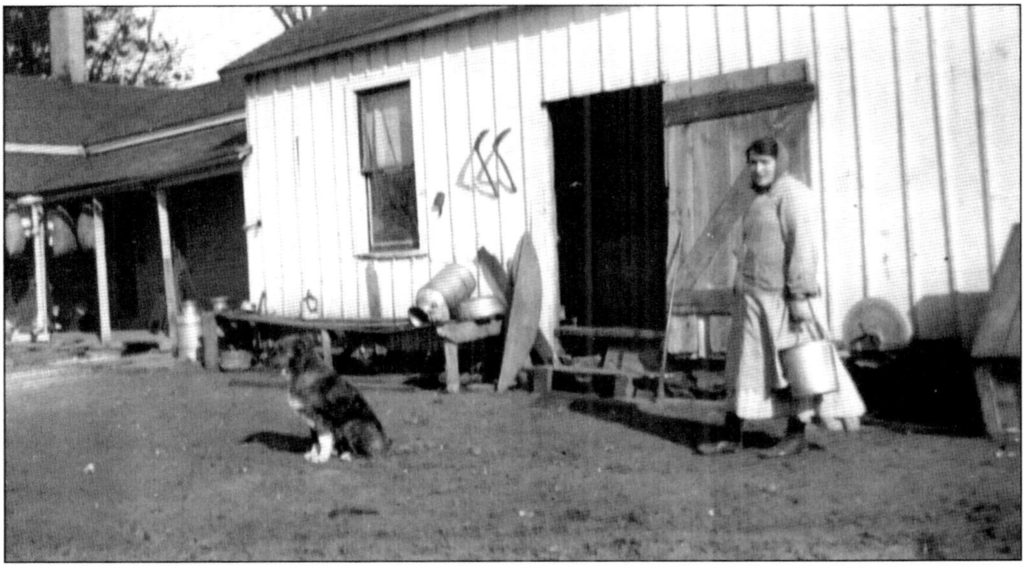

Farm women who wanted to contribute to the family's income often raised and sold chickens, but many farmers' wives had little other contact with the outside world, instead tending gardens and preparing food and sewing clothing for the family. In 1887, Alura Collins Hollister of Mukwonago said that a farm woman often "forgets that life was not given to her simply to bear children and save money." (WHS.)

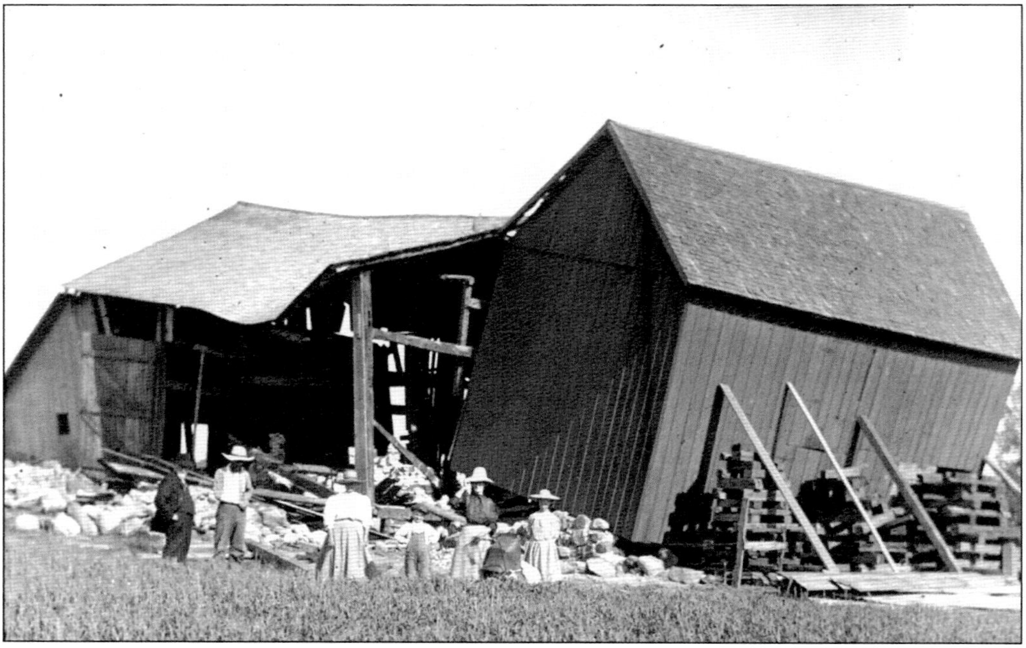

The Tratt farm, pictured, found itself the victim of destructive winds. This particular barn was blown off its base. Farmers knew that high-speed winds and tornadoes were frequent in Wisconsin, so they often took special care to secure their buildings. Unfortunately, they were not always successful. Frank W. Tratt operated this farm in Whitewater Township. (Irvin L. Young Memorial Library; Fran Achen Collection.)

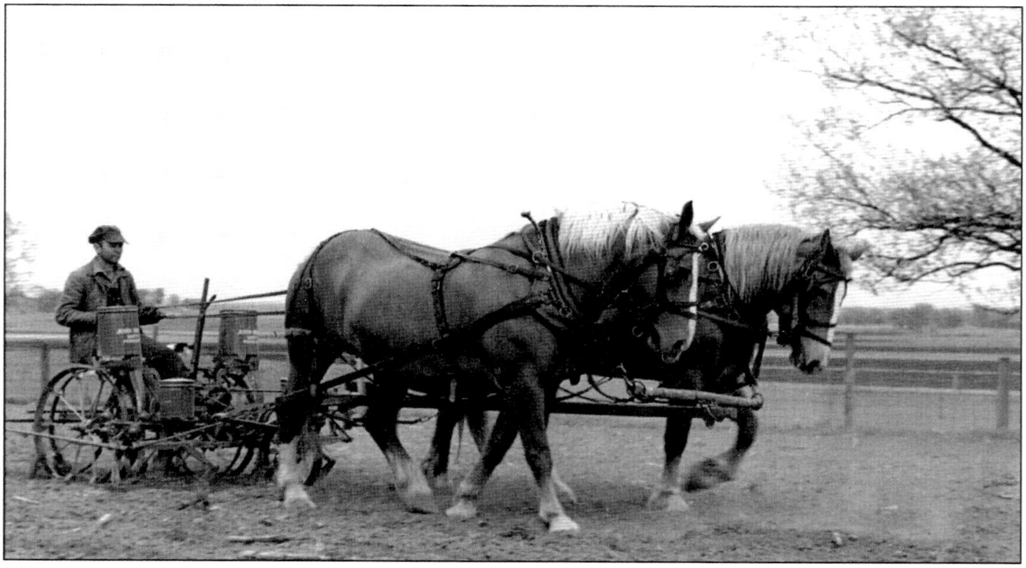

Horse-drawn farm implements like the seeder seen here were used well into the 20th century on Whitewater farms. Mechanized equipment was certainly much more useful, but this increased utility came at a much higher price. Small family farmers usually could not afford the latest equipment, instead continuing to use their old, reliable, horse-powered machines. (Irvin L. Young Memorial Library; Fran Achen Collection.)

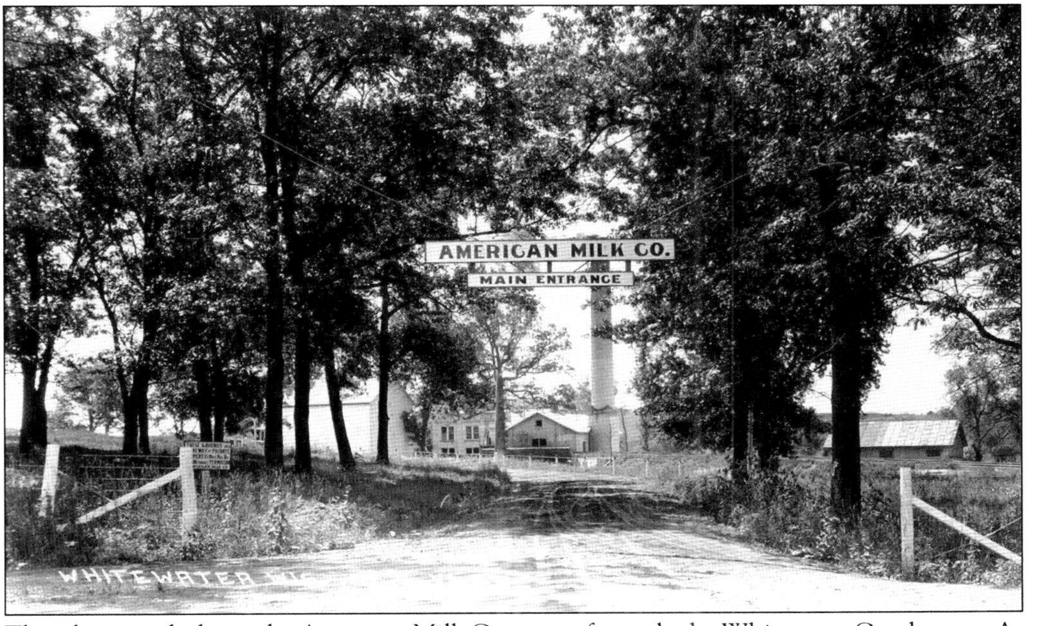

This photograph shows the American Milk Company, formerly the Whitewater Condensery. At condenseries such as this, cow's milk was steamed and processed to remove water. The thickened milk was then put through a homogenizer to break up the butter and fat. The canned milk could keep for years without refrigeration. Consumers simply had to add the proper amount of water before consuming the milk. (WHS.)

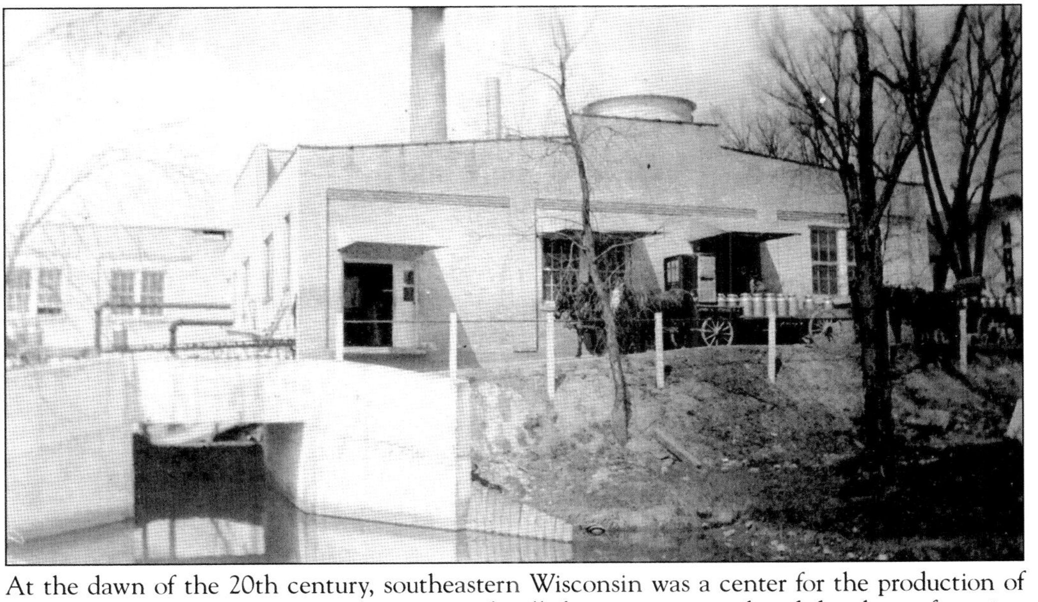

At the dawn of the 20th century, southeastern Wisconsin was a center for the production of American cheese. However, condenseries and milk factories soon replaced the cheese factories. The American Milk Company was taken over by Hawthorn Mellody in 1944. In 1950, Kenosha, Racine, and Walworth Counties shipped more than 311 million pounds of milk out of state. (Irvin L. Young Memorial Library; Fran Achen Collection.)

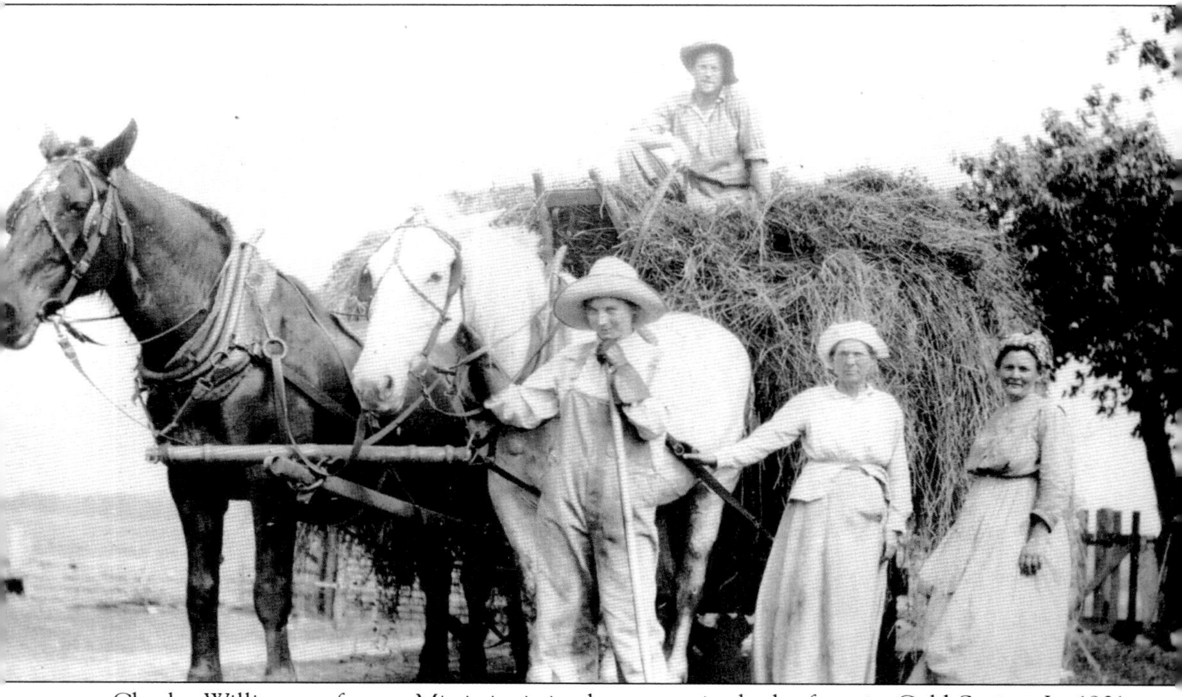

Charles Williams, a former Mississippi riverboat captain, had a farm in Cold Spring. In 1901, Williams assumed ownership of the farm from his wife's family. It was called the Three Oaks Dairy and Poultry, and turkeys and cows were their main livestock. Pictured harvesting hay are Charles W. Williams on top of wagon, and from left to right, Charles's sister Maud; his mother, Josephine Williams; and his wife, Nettie Henderson Williams. (Henderson family.)

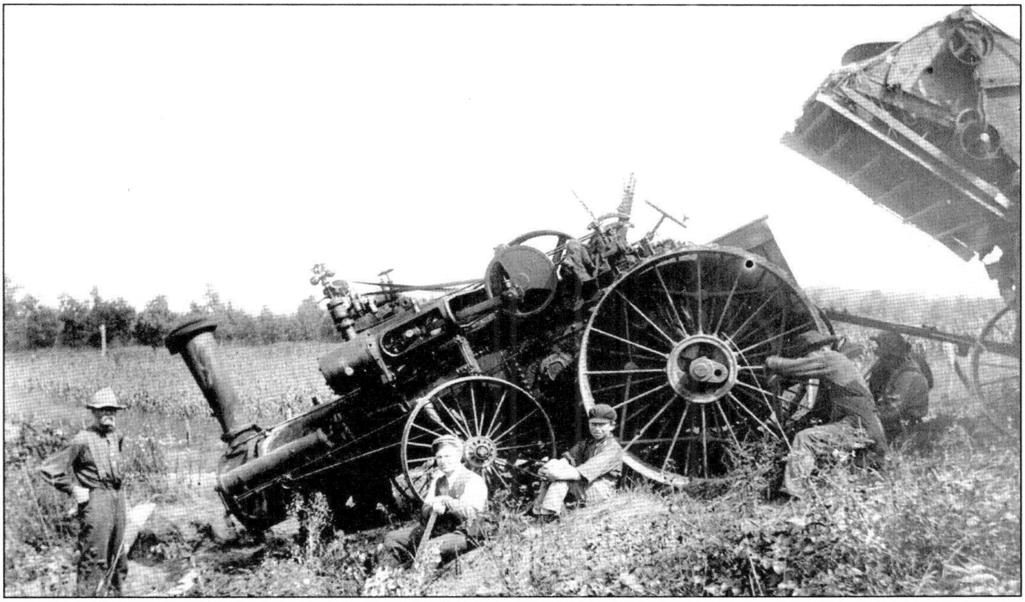

Charles Sprackling, at left, is pictured here at the Roach farm in LaGrange, as he prepares to fix this steam tractor that has run into a ditch. Few farmers could afford these machines, and even those who could were wary of them, as they were difficult to repair. Early steam tractors were primarily used for heavy hauling, plowing, threshing, and packing down soil. (Irvin L. Young Memorial Library; Fran Achen Collection.)

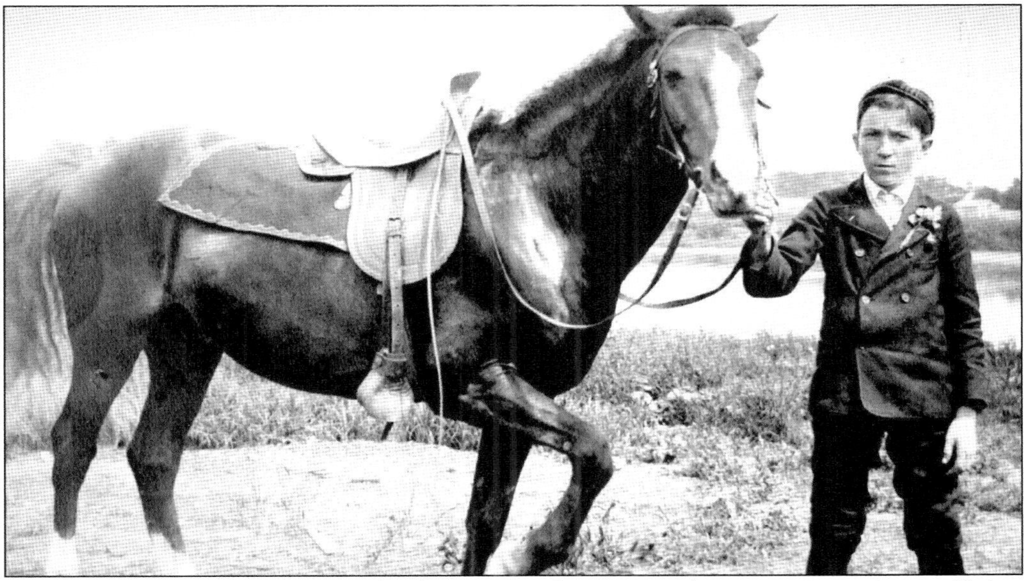

Danny Belknap is seen here with his horse around 1900 in Whitewater. Danny's grandfather Daniel D. Belknap was a well-known physician in Whitewater who had an office just north of the Esterly Company. Before the automobile was commonly used, families kept horses within the city limits. For small towns like Whitewater, the switch to automobiles was slow, and horses were common in town well into the 1920s. (Irvin L. Young Memorial Library; Fran Achen Collection.)

AFTERWORD

There is no simple way to describe why some towns fail and some do not. One lesson to be learned from this work is that economic decline, even on a local basis, is the result of many factors, both controllable and uncontrollable. The panic of 1893 was devastating, but Whitewater citizens' local policies and actions were perhaps just as harmful. The ideal town will be prepared, flexible, and responsive enough to withstand a depression. Thanks to human error, shortsightedness, and fate, Whitewater was not able to maintain its previous level of success.

Today Whitewater is a successful, growing city. It boasts an excellent university, as well as many successful businesses and factories. But Whitewater's great slowdown perhaps prevented the town from becoming even more. A unique combination of circumstances allowed Whitewater to boom from 1850 to 1890, but a similarly unique combination caused the town to bust. Perhaps it was the invisible hand of fate that caused Whitewater's fortunes to turn so rapidly, as so many different factors seemed to conspire against the town. The events of 1893 constituted the most difficult test that the city of Whitewater ever faced. While the town's response as a whole was flawed, some individuals put their financial resources and acumen to work on rebuilding the town. Their fighting spirit is what kept the town alive.

SELECTED BIBLIOGRAPHY

Bensel, Richard Franklin. *The Political Economy of American Industrialization, 1877-1900.* Cambridge: Cambridge University Press, 2000.

Brands, H. W. *The Reckless Decade*. New York: St. Martin's Press, 1995.

Cravath, Prosper. *Early Annals of Whitewater.* Whitewater, WI: 1906.

Ell Torrance Papers, Box 80-82, Minnesota Historical Society.

Flower, Frank A., et al. State of Wisconsin. *Public document no. 9: third biennial report of the bureau of labor and industrial statistics. Wisconsin.* 1889.

Hoffman, Charles. "The Depression of the Nineties." *The Journal of Economic History* 12, No. 4 (November 1930): 181–185.

Hunt, Robert. *Law and Locomotives: The Impact of the Railroad on Wisconsin Law in the Nineteenth Century.* Madison, WI: the State Historical Society of Wisconsin, 1958.

Kirkland, Edward Chase. *Industry Comes of Age: Business, Labor and Public Policy, 1860-1897.* Chicago: Quadrangle Paperbacks, 1961.

Leavitt, Charles. "Transportation and the Livestock Industry of the Middle West to 1860." *Agricultural History* 8, No. 1 (January 1934): 20–33.

"Notes on Municipal Government." *Annals of the American Academy of Political and Social Science* 12 (November 1898): 112–125.

Noyes, Alexander. "The Banks and the Panic of 1893." *Political Science Quarterly* IX (1894): 12–30.

Rikoon, J. Sanford. *Threshing in the Midwest, 1820-1940.* Bloomington, IN: Indiana University Press, 1988.

Smith, Page. *A People's History of the Post-Reconstruction Era Volume 6: The Rise of Industrial America.* New York: McGraw-Hill Book Company, 1984.

Smith, William R. et al. "Sketch of Whitewater." *Governor's message and accompanying documents for the year 1857.* Madison, WI: 1858.

Whitbeck, R. H. *The geography and economic development of southeastern Wisconsin.* Madison, WI: State of Wisconsin, 1921.